Geometry and Computing

T0223365

For further volumes:
www.springer.com/series/7580

Malcolm Sabin

Analysis and Design
of Univariate Subdivision
Schemes

With 76 Figures

 Springer

Malcolm Sabin
Numerical Geometry Ltd,
John Amner Close 19
CB6 1DT Ely
United Kingdom
malcolm@geometry.demon.co.uk

ISSN 1866-6795   e-ISSN 1866-6809
ISBN 978-3-642-26449-8   ISBN 978-3-642-13648-1(eBook)
DOI 10.1007/978-3-642-13648-1
Springer Heidelberg Dordrecht London New York

Mathematics Subject Classification (2010): 53A04, 65D05, 65D07, 65D10, 65D17, 65D18, 68U05, 68U07

Cover design: deblik, Berlin

Printed on acid-free paper

Springer is part of Springer Science+Business Media (www.springer.com)

Preface

'Subdivision' is a way of representing smooth shapes in a computer. A curve or surface (both of which contain an infinite number of points) is described in terms of two objects. One object is a sequence of vertices, which we visualise as a polygon, for curves, or a network of vertices, which we visualise by drawing the edges or faces of the network, for surfaces.

The other object is a set of rules for making denser sequences or networks. When applied repeatedly, the denser and denser sequences are claimed to converge to a limit, which is the curve or surface that we want to represent.

This book focusses on curves, because the theory for that is complete enough that a book claiming that our understanding is complete is exactly what is needed to stimulate research proving that claim wrong. Also because there are already a number of good books on subdivision surfaces.

The way in which the limit curve relates to the polygon, and a lot of interesting properties of the limit curve, depend on the set of rules, and this book is about how one can deduce those properties from the set of rules, and how one can then use that understanding to construct rules which give the properties that one wants.

This book therefore has four main parts. First are a set of 'Prependices' which are potted descriptions of little bits of mathematics which turn out to be useful background. These can be skipped at first reading, or by those who know the material anyway.

Then a chapter introducing the concepts for the reader who has not encountered subdivision curves and surfaces before; and how the rules are described in ways that we can apply mathematical arguments to.

Third a set of chapters dealing with the ways that we can analyse properties of the limit curves in terms of the rules, followed by a shorter set of chapters suggesting how we can work the analyses in reverse, to design rules which will give the schemes that result from them desired properties.

Where a chapter introduces techniques worth trying out, it also includes a few exercises to help the reader check that they have indeed been learned.

Finally come some ideas about efficient implementation and some appendices tidying up material best kept out of the way of the main flow of ideas, including some topics open for research, a short account of the development of the subject so far, and a bibliography of research papers.

I would like to thank the following people for their help and forbearance during the writing of this book: first, of course, my wife for her patience when there were other things she would have liked me to be doing, but also David Levin for his patient explanation of z-transforms to me, Adi Levin for showing me how the joint spectral radius worked and Carl de Boor for showing me how to say what I meant about eigenanalysis in terms of invariant subspaces and for the idea of Theorem 4.

Then there are the colleagues who have read early proofs, the referees whose comments have triggered significant improvements, and the staff at Springer who have tolerated my views as to how the book should look.

Thank you all.

July 2010, Ely

Table of Contents

Part II. Dramatis Personae

Part III. Analyses

Part V. Implementation

Part VI. Appendices

Notation

Within this book, lower case letters are used to denote scalars, upper case letters more complicated things, such as vectors, matrices, sequences, etc. with the exception that in chapters related to eigenanalysis corresponding objects use the same letter, in uppercase before multiplication by a matrix, in lowercase after, in order to stress the correspondence.

Band matrices have the region outside the band left blank, rather than being filled with zeroes, or labelled as being zero.

Italics are used for stressing phrases, or for quotations (such as the echoing of exercise questions within the worked solutions). Bold font is used for words or phrases being defined by the text around them.

Superscripts are always either powers or an indicator of degree. Apart from the more complicated expressions found in Hölder continuity, they are integers. Subscripts indicate distance along a sequence and need not be integral. When writing code the reader is advised to make explicit the mapping from the subscripts in this book to subscripts into arrays referenced in that code.

Part I. Prependices

This section of the book provides brief introductions to a number of background pieces of mathematics which are relevant, directly or indirectly, to the analysis of subdivision curves.

These are:

- Functions and curves
- Differences
- B-splines
- Eigenanalysis
- Enclosures
- The Hölder measure of derivative continuity
- Matrix norms
- The joint spectral radius of two matrices
- Radix notation
- z-transforms

The polymath-ematician and the impatient should turn directly to the main body of the book, on page 49. Those who want or need to be better prepared with a little revision of mathematical matters learnt early and easily forgotten, should first skim through these introductory pages, noting what is here, so that, when they find the need for a reminder, they will know where to look for it.

These introductions are not, of course, full descriptions of their topics: they are mainly lists of the results relevant to the main body of the book.

1. Functions and Curves

The idea of a function is a familiar one. The value of the function (the ordinate) depends on the value of the argument (the abscissa) of the function.

Graphing a function plots the ordinate vertically against the abscissa horizontally, which gives a visible curve.

The coefficients of the function can be given geometric meaning.

A graph of $y = ax^2 + bx + c$

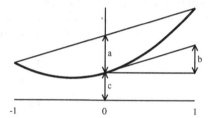

However, when we wish to describe shapes of objects, it is rather important that we should be able to express them in a way which reflects the fact that shapes are invariant under solid body transforms, such as translations and rotations. This is usually done by having two or three ordinates, all functions of the same abscissa. Each function is a sum of terms where each term is the product of a **coefficient** and a **basis function**, taken from a set common to all the coordinates.

$$x = \Sigma_i x_i f_i(t)$$
$$y = \Sigma_i y_i f_i(t)$$
$$z = \Sigma_i z_i f_i(t)$$

We can then think of the coefficients corresponding to a given basis function as forming a vector coefficient.

$$P = \Sigma_i C_i f_i(t)$$

These vector coefficients then typically become interpretable as either points, or displacements[1] and applying the appropriate solid body (or even

---

[1] The distinction between points and displacements seems pedantic, but is actually useful in defensive coding. Points and displacements transform differently under translations.

M. Sabin, *Analysis and Design of Univariate Subdivision Schemes*, Geometry and Computing 6, 5
DOI 10.1007/978-3-642-13648-1_1, © Springer-Verlag Berlin Heidelberg 2010

affine) transformation to these coefficients has the effect of applying that transformation to the entire curve.

A plot of $P = At^2 + Bt + C$. C is a point and A and B displacement vectors.

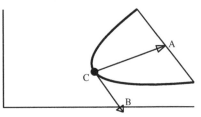

For polynomials in monomial form, the constant coefficient is a point and all the others are displacements.

For shape design purposes it is usually far more convenient to use a basis where the basis functions sum to 1. All the coefficients then transform as points, and we call them **control points**. For polynomials this can be achieved by using the **Bézier** basis, which is a special case of the **B-spline basis** which will be encountered shortly.

For quadratics, the Bézier basis functions are

$$b_0(t) = (1 - t)^2$$
$$b_1(t) = 2t(1 - t)$$
$$b_2(t) = t^2$$

which maps the interval $0 \le t \le 1$ into a piece of curve.

A plot of $P(t) = \Sigma P_i \, b_i(t)$. P_0, P_1 and P_2 are all points.

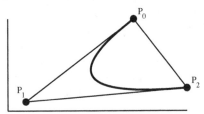

1.1 Summary

(i)   When representing shapes of physical objects it is convenient to use position-valued functions where each of three coordinates is a function of a common abscissa, or **parameter**.

(ii)   If this function is expressed as the weighted mean of a set of point coefficients, each multiplied by a scalar-valued basis function, where the set of basis functions sums to unity everywhere, the complete curve can be transformed by transforming the points.

(iii) The way that the shape of the curve is related to the positions of its defining points depends on properties of the basis functions.

2. Differences

Differences are a fundamental idea in approximation theory. They are encountered in tables of functions as a way of supporting the production, at look-up time, by linear interpolation, of values in between the actually tabulated values. Their theory was developed at a time when creation of tables of functions was an important activity supporting navigation. The theory is now more important than that particular application.

2.1 First Differences

Consider a sequence, V, of values, V_i, $i \in \mathbb{Z}$

The sequence of **differences**, δV, is formed from exactly the differences between each value and the next in sequence, $V_i - V_{i-1}$. There is only one concept, but there are three ways of assigning a subscript to a particular difference value.

In **forward differencing** we let $\delta V_i = V_i - V_{i-1}$: in **backward differencing** we let $\delta V_{i-1} = V_i - V_{i-1}$. In **central differencing** we say that a particular difference is associated with the interval between two of the original values, and that therefore $\delta V_{i-1/2} = V_i - V_{i-1}$. While this is less convenient for computation (how many programming languages allow for half-integer valued subscripts ?), it reflects the semantics of the sequence better, and we will use it as our convention.

Differences are not necessarily just scalars. The first differences of the vertices of a polygon are vectors along the directed edges. Clearly it is sensible to associate these with abscissae halfway between the vertices.

2.2 Higher Differences

Differencing is an operation on a sequence which gives another sequence. We can therefore apply it again to that result sequence. This gives the sequence of **second differences**. $\delta(\delta V)$ is naturally denoted by $\delta^2 V$, and δ^2 is called the **second difference operator**. If we use the central difference convention we get

M. Sabin, *Analysis and Design of Univariate Subdivision Schemes*, Geometry and Computing 6, 7
DOI 10.1007/978-3-642-13648-1_2, © Springer-Verlag Berlin Heidelberg 2010

$$\delta^2 V_i = (\delta^2 V)_i = V_{i+1} - 2V_i + V_{i-1}$$

Still higher differences are defined in exactly the same way.

2.3 Differences of Polynomials

Consider a polynomial of degree zero. If this is evaluated at integer values we get a sequence of values all the same, and the difference sequence is all zeroes. We say that the differencing operator δ **annihilates** polynomials of degree zero.

If the polynomial is of degree 1, then the values in the sequence vary linearly, and the first differences are all the same. These are in turn annihilated by the application of a second differencing operator. Thus δ^2 annihilates polynomials of degree 1.

In general δ^d annihilates polynomials of degree $d - 1$.

x	x^2	$\delta(x^2)$	$\delta^2(x^2)$	$\delta^3(x^2)$
0	0			
		1		
1	1		2	
		3		0
2	4		2	
		5		0
3	9		2	
		7		0
4	16		2	
		9		0
5	25		2	
		11		
6	36			

Central Differences of x^2

2.4 Divided Differences

If we have an abscissa associated with the ordinate values in the sequence, it is sensible to divide the first differences of the ordinates by the first differences of the abscissae, to give the mean slope of the chords joining the data points represented by the values in the two sequences. Such a ratio is called the **divided difference**.

$$\delta y_{i-1/2} = \frac{y_i - y_{i-1}}{x_i - x_{i-1}}$$

The significance of this is that the derivative of a function is defined as the limit of the divided difference of samples as the gaps in abscissa tend to zero.

$$\frac{dy}{dx} = \lim_{\delta x \to 0} \frac{\delta y}{\delta x}$$

If the data is indeed sampled at regular intervals from a polynomial of degree d, then the dth divided difference, which is a constant sequence, has as the value of each element the value of the dth derivative.

x	x^2	$\delta(x^2)$	$\delta^2(x^2)$	$\delta^3(x^2)$		x	x^2	$\frac{\delta(x^2)}{2}$	$\frac{\delta^2(x^2)}{4}$	$\frac{\delta^3(x^2)}{8}$
0	0					0	0			
		1						2		
1	1		2			2	4		2	
		3		0				6		0
2	4		2			4	16		2	
		5		0				10		0
3	9		2			6	36		2	
		7		0				14		0
4	16		2			8	64		2	
		9		0				18		0
5	25		2			10	100		2	
		11						22		
6	36					12	144			

Central Divided Differences: (left) at unit intervals, (right) at intervals of 2. The values of divided differences are no longer scaled by the sampling density.

The appropriate denominators are straightforward to see in the uniform case, where we can say that the divided differences come half-way between the original values, and so we have well-defined abscissae associated with the difference values, and it is clear what is meant by saying that the second divided differences are divided by the difference in abscissa of the first divided differences. In the non-uniform context the convention often used is that the second divided differences are the differences of the first divided differences divided by half of the total width of the set of values taking part in the divided difference. In the uniform subdivision context we do not need that complication.

2.5 Summary

(i)   Differences are determined by subtracting each entry in a sequence from
        the next.

(ii)   They can be expressed in terms of forward, central or backward differ-
        ences, depending on which original entry a specific difference value is
        associated with. For our purposes the central difference is most appro-
        priate.

(iii)   Higher differences (second, third, ...) are just differences of the differ-
        ences.

(iv)   If the values in the sequence are taken from a polynomial, a high enough
        difference will become a sequence of zeroes.

(v)   Divided differences are defined as the differences, divided by the differ-
        ence in abscissa.

(vi)   Higher divided differences are just divided differences of divided differ-
        ences, with the abscissae of the lower differences chosen sensibly.

(vii)   If these are of samples from a smooth curve they converge towards the
        corresponding derivatives as the sampling interval reduces.

3. B-Splines

3.1 Definition

A **spline** is a piecewise polynomial[2] whose pieces meet with continuity as high as possible given the degree. The abscissa values at which consecutive pieces meet are called the **knots**. A **B-spline** is a spline expressed with respect to a particular basis, in which the basis functions are each non-zero over as small a number of consecutive pieces as possible, given the degree and the continuity, and the basis functions sum to unity.

If the pieces are all of equal length in abscissa, all the B-spline basis functions are just translates of the same basic function, which is typically called *the* basis function of the given degree. The resulting curves are then called **equal interval B-splines** or **uniform B-splines**. We shall use the shorter term 'B-splines' here, despite the fact that it is not strictly accurate.

The characterisation of B-splines that we shall use here is that the B-spline basis function of degree n has the following properties

- It consists of pieces of polynomial of degree n.
- Its pieces meet (at integer values of abscissa) with continuity of $(n-1)^{th}$ derivative.
- It is nonzero only within a sequence of $n+1$ consecutive intervals.
- Its integer translates sum to 1 identically.
- It is non-negative[3].

There are well over a dozen ways in which the function with these properties can be described.

---

[2] There are useful generalizations to piecewise functions where the pieces are smooth functions other than polynomials (for example, trigonometric or exponential splines), but the polynomial definition is most relevant at this point.

[3] This property can actually be derived from the first four, which are a sufficient characterisation.

M. Sabin, *Analysis and Design of Univariate Subdivision Schemes*, Geometry and Computing 6,
DOI 10.1007/978-3-642-13648-1_3, © Springer-Verlag Berlin Heidelberg 2010

One of the nicest is as the variation of the cross-sectional 'area' of a unit $(n+1)$-dimensional cube when cut by a n-dimensional 'plane' perpendicular to the diagonal, as a function of the distance along that diagonal at which the plane is placed.

The five properties above are all evident from the figure on the right, which shows the case $n = 2$, taken together with the fact that the cross-sectional area around the leftmost point of the box is given by the truncated n^{th} power.

We can use these functions to describe curves by using the translates as basis functions for a point-valued parametric curve

$$P(t) = \sum_j b_j^n(t) P_j$$

where the coefficients P_j are control points and $b_j^n(t) = b^n(t - j)$ are the B-spline basis functions of degree n.

Because the basis functions sum to unity, the coefficients transform as points, and are called **control points** and they are typically visualised by drawing the polygon which joins them in sequence.

3.2 Derivative Properties

Clearly the first derivative of the B-spline basis function must have the properties that

• It consists of polynomial pieces of degree $n - 1$, because each piece is the derivative of a polynomial of degree n .

• Its pieces meet with continuity of $(n - 2)^{th}$ derivative, because the discontinuities of the original are of the $(n - 1)^{th}$.

• It is non-zero only in a sequence of $n + 1$ intervals because the function of which it is a derivative is identically zero outside that range .

From these we see that it must be a linear combination of exactly two B-spline basis functions of degree $n - 1$, and in fact it is exactly the difference of two consecutive basis functions of degree $n - 1$.

$$b_j'^n(t) = b_j^{n-1}(t) - b_{j-1}^{n-1}(t)$$

The first derivative of a B-spline curve is a vector-valued function given by

$$P'(t) = \sum_j b_j'^n(t) P_j$$
$$= \sum_j (b_j^{n-1}(t) - b_{j-1}^{n-1}(t)) P_j$$
$$= \sum_j b_j^{n-1}(t)[P_j - P_{j-1}].$$

This is exactly a B-spline of one degree lower, with the first differences of the original control points in the rôle of control vectors.

The second derivative vector is a B-spline of two degrees lower, with the second differences of the original control points in the rôle of control vectors, and higher derivatives have analogous properties, as far as the degree of the B-spline permits.

When we take the nth derivative of a degree n spline, we get a piecewise constant. If we try to take the $(n+1)^{th}$ derivative we get zero within all of the pieces, and, in order to be able to recover the nth derivative by integration, there have to be Dirac delta-functions (infinite spikes of zero width) at all of the knots where the pieces meet.

3.3 Construction

We can use this property of derivatives in reverse to construct B-splines of arbitrary degree.

The B-spline b^0 of degree zero is just the function

$$b^0(x) = \begin{cases} 0 & x < 0 \\ 1 & 0 \le x < 1 \\ 0 & x \ge 0 \end{cases}$$

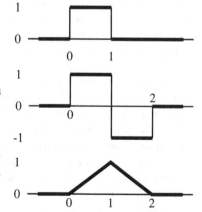

If we subtract the unit translate of this from the zero translate we get a function

and this is the first derivative of the degree 1 B-spline, which can be constructed by integration.

Higher degree B-splines can be constructed explicitly (using, for example, the Bézier basis for each span of the function) by recursion on degree, applying the same simple recipe:

- shift

- subtract

- integrate

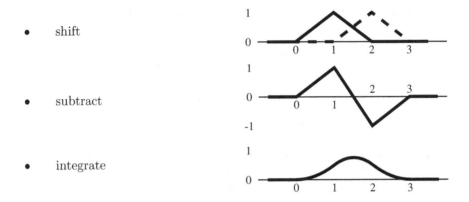

$$b^{n+1}(t) = \int_0^t (b^n(s) - b^n(s-1))\, ds.$$

This process can also be viewed as convolution with the 0-degree box-spline, expressed by integration by parts.

$$b^{n+1}(t) = \int b^n(s) b^0(t-s)\, ds.$$

3.4 Refinement

A standard operation on piecewise polynomial curves is that of **knot insertion** where we pretend that there are additional **knots** (junctions between pieces) at which the discontinuity of nth derivative happens to be of zero magnitude.

In the B-spline context we need the pieces after knot insertion all to be of the same length, and so the knot insertion which maps B-splines into B-splines inserts the new knots at the half-integers. The abscissa is then scaled so that the pieces are again of unit length.

The refined curve has a refined basis, and needs a corresponding sequence of control points. In fact it is the process of finding these new control points which is actually called **knot insertion** or **subdivision**. If we ignore end-conditions, which is a sensible way to start, there are twice as many of them. We can determine them by looking at the way in which a coarse basis function can be expressed in terms of the refined ones. Let the coarse basis be $b^n_\tau(t)$ and the finer one $\bar{b}^n_\tau(t)$ where the subscripts τ indicate the position in abscissa space of the central maximum[4] of the particular basis function, and the superscript n the degree of the functions. Consider first B-splines of degree zero.

---

[4] The central convention will be seen later to help in exploitation of the symmetry properties of these functions.

Clearly

$$b_\tau^0(t) = \bar{b}_{\tau-1/4}(t) + \bar{b}_{\tau+1/4}(t).$$

Hence

$$P(t) = \sum_\tau b_\tau^0 P_\tau$$

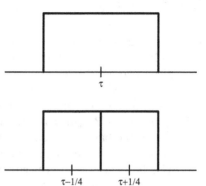

$$= \sum_\tau \left(\bar{b}_{\tau-1/4}(t) + \bar{b}_{\tau+1/4}(t) \right) P_\tau$$

$$= \sum_\tau \bar{b}_{\tau-1/4}(t) P_\tau + \bar{b}_{\tau+1/4}(t) P_\tau$$

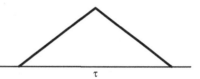

This will be equal to $\sum_{\bar\tau} \bar{b}_{\bar\tau} \bar{P}_{\bar\tau}$ if

$$\bar{P}_{\tau-1/4} = P_\tau$$
$$\bar{P}_{\tau+1/4} = P_\tau$$

Thus the refinement process for degree zero B-splines is merely the taking of each original control point twice.

For degree one B-splines, we have the relation between the new and old basis functions that

$$b_\tau^1(t) = \left(\bar{b}_{\tau-1/2} + 2\bar{b}_\tau + \bar{b}_{\tau+1/2} \right) / 2$$

which leads to the relation between old and new control points that

$$\bar{P}_{\tau-1/2} = \left(P_\tau + P_{\tau-1} \right) / 2$$
$$\bar{P}_\tau = P_\tau$$

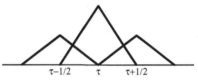

For higher degree B-splines, degree n say, we can determine the new control points by taking means of consecutive control points for degree $n - 1$. This is called the **Lane-Riesenfeld** construction.

In fact the degree=1 refinement construction is exactly taking the degree=0 construction (taking each control point twice) and taking means of consecutive control points.

We can go further, by inventing a degree=-1 construction which doubles the coordinates of all the control points and intersperses them with new points with zero coordinates. The degree=0 construction is then recovered by applying the Lane-Riesenfeld construction.

3.5 Summary

(i)   A spline of degree d is a piecewise function (or parametric curve) with
polynomial pieces of degree d meeting with continuity of $d - 1^{th}$ derivative. These pieces can be called **spans**.

(ii) The places where pieces meet are called **knots**, and in the uniform case
these are equally spaced in abscissa.

(iii) A **uniform B-spline** is a spline defined in terms of the **B-spline basis**,
where the basis functions are all translates of the same function, which
is zero outside an abscissa interval $d + 1$ pieces wide. The coefficients in
this basis are called **control points**.

(iv) This sequence of control points is typically visualised by joining them
by straight lines, and so it is called the **control polygon**.

(v)   This basis has the nice property that derivatives are also representable
as B-splines.

(vi) The refinement by knot insertion at the midpoints of the spans is given
by a particularly simple geometric construction on the control polygon.

4. Eigenfactorisation

4.1 Definition

Eigenfactorisation is the splitting of a square matrix, M, into three factors

$$M = C L R$$

such that $RC = I$ and L has as few as possible off-diagonal non-zero entries.

The purpose of this factorisation is to help understand what happens when a vector is multiplied by the same matrix repeatedly, as happens in subdivision.

If L is diagonal, with no off-diagonal non-zeroes at all, then the columns of C are called **eigenvectors** and the diagonal values in L are called **eigenvalues**. In the more general case, when L is not completely diagonal, they are supposed to be called **generalised eigenvectors**, but we shall use the shorter phrase.

In fact in conventional eigenanalysis the focus is on the columns rather than the rows. (The row eigenvectors, R, are the column eigenvectors of the transpose of the matrix M.) The distinction is made between **right-eigenvectors** and **left-eigenvectors** depending on whether the object in question is multiplied by a matrix on the left or on the right. In subdivision they each have distinct rôles, and we shall use the more evocative terms **eigencolumns** and **eigenrows**.

We consider first the "normal" case, when there are no off-diagonal non-zero entries in L, and then the more general case.

4.2 Uniqueness

The factorisation is not unique, for three reasons.

(i) Permutation

Suppose that we have any permutation matrix, P. Then $P^{-1}P = I$ and we can rewrite the defining equation as

M. Sabin, *Analysis and Design of Univariate Subdivision Schemes*, Geometry and Computing 6,   17
DOI 10.1007/978-3-642-13648-1_4, © Springer-Verlag Berlin Heidelberg 2010

$$M = CILIR$$
$$= C(P^{-1}P)L(P^{-1}P)R$$
$$= CP^{-1}PLP^{-1}PR$$
$$= C'L'R'$$

C' is a matrix containing a permutation of the columns of C, R' is a matrix containing the same permutation of the rows of R, and L' is a matrix whose diagonal elements are permuted in the same way, and whose non-diagonal elements are moved to different places.

The correspondence between a particular diagonal element and the associated column and row is not affected, and so we are free to take the triples of eigenvalue, column and row in whatever sequence we like. In this book we call such a triple an **eigencomponent**.

There is a widely adopted convention that the largest eigenvalue should be at the top left hand corner of L and the remainder should be in decreasing sequence of magnitude. In part of the analysis in the body of the book we have a good reason for using a different convention, but we remain conventional until that good reason appears.

The eigenvalue with the largest magnitude is called the **dominant** eigenvalue, the next the **subdominant**, and so on.

(ii)   Scaling

Let each column C_j in C be scaled by an independent non-zero scalar s_j. Then provided that each row R_j is scaled inversely by the factor, the product CLR remains unchanged.

There is a widely adopted convention that the columns should be scaled so that the sum of the squares of the elements of a column should be unity.

Again, we have a good reason for sometimes using a different convention, that if the sum of the elements of a row is not zero, the row should be scaled so that that sum is unity, the column being scaled inversely.

(iii)   Rotation

In the particular case where two diagonal entries of L have the same value, and there is no off-diagonal linking them, then any independent pair of linear combinations of the two columns may be used instead of the originals.

There is no standard convention here; neither do we need anything special, but this property will be used later.

In fact these are just particular instances of a similarity transform, in which the matrix is pre-multiplied by a non-singular matrix and post-multiplied by its inverse

$$TMT^{-1} = TCLRT^{-1}$$
$$= TCT^{-1}TLT^{-1}TRT^{-1}.$$

If L is diagonal, TLT^{-1} is also diagonal, containing the same values. We can think of this process as expressing the matrix with respect to a different coordinate system.

4.3 Properties of Eigenvectors

What happens when an eigenvector is multiplied by a matrix ?

We consider first the case where L is a diagonal matrix, so that there are no off-diagonal entries. The more general case will be considered on page 22 below.

$$MC_j = CLRC_j$$
$$= CL\Delta_j \text{ where } \Delta_j \text{ is a column vector with a 1 in the } j^{th} \text{ place}$$
$$\text{and zeroes elsewhere.}$$
$$= C\lambda_j\Delta_j \text{ where } \lambda_j \text{ is the } j^{th} \text{ diagonal element}$$
$$= \lambda_j C_j$$

Thus the effect of multiplication by the matrix on an eigenvector is to multiply that vector by the corresponding eigenvalue.

$$MC_j = \lambda_j C_j$$

4.3.1 What happens when a general vector is multiplied by a matrix?

Assuming that the matrix C is of full rank (which is a safe assumption because it has an inverse, R), we can write a general vector, V, as a weighted sum of the columns of C

$$V = \Sigma_j \alpha_j C_j$$

In fact the weights α_j, which can be thought of as a column vector A, can be computed by premultiplying V by R

$$RV = RCA$$
$$= A$$

but actually computing them is not important at this point of the argument.

$$MV = \Sigma_j \alpha_j M C_j$$
$$= \Sigma_j \alpha_j \lambda_j C_j$$
$$= \Sigma_j \alpha'_j C_j$$

so that the product is also a weighted sum of the eigenvectors, but with the weights multiplied by the eigenvalues.

$$\alpha'_j = \alpha_j \lambda_j$$

Thus the new vector is again a weighted mean of the eigenvectors, but with the weights each multiplied by the corresponding eigenvalue.

4.3.2 What happens when a general vector is multiplied by a matrix repeatedly?

At every multiplication the weights get multiplied by the eigenvalues. Thus as the number of multiplications increases, the contribution of the eigenvector with the largest eigenvalue gets to be more and more dominant. This is why that eigencomponent is called the dominant one.

If, however, the original vector happened to be orthogonal to the dominant eigenrow, then there would be nothing of that component to grow relative to the others, and in those circumstances the second eigencomponent will dominate.

4.4 Calculating Eigencomponents

This is non-trivial. We can see this by noting that the property $MC_j = \lambda_j C_j$ can be rewritten as

$$[M - \lambda_j I]C_j = 0$$

The matrix $[M - \lambda_j I]$ must be of reduced rank to give a zero result when it multiplies a non-zero vector, and it therefore has a zero determinant.

Thus λ_j is a root of the equation, polynomial in λ, $\det(M - \lambda I) = 0$ and computing the eigenvalues is equivalent to finding the roots of that polynomial, which is called the **characteristic polynomial**.

Galois proved that this is non-trivial. If the size of M is greater than 4×4, then there is no algebraic closed form, and if the size of M is greater than 2×2 there is no practical closed form.

Any algorithm for computing eigenvalues must therefore be iterative. The iteration might be hidden inside a polynomial solver or explicit (as in the QR algorithm) but it will always be there, unless some other information about the matrix can be exploited.

Once an eigenvector is known, computing the corresponding eigenvalue can be done by just multiplying the vector by the matrix and seeing what the scaling factor is. Also, once an eigenvalue is known, determining the corresponding eigenvector is just a question of solving a linear system.

Block structure

A particular example of information which can be exploited is when the matrix M is of block triangular structure.

Let M be formed from the three sub-matrices

$$M = \begin{bmatrix} D & \\ E & F \end{bmatrix}$$

where D and F are square.

Then the vector $V = \begin{bmatrix} 0 \\ Y \end{bmatrix}$ where Y is an eigenvector of F with eigenvalue λ, will be an eigenvector of M with the same eigenvalue.

$$\begin{aligned} MV &= \begin{bmatrix} D & 0 \\ E & F \end{bmatrix} \begin{bmatrix} 0 \\ Y \end{bmatrix} \\ &= \begin{bmatrix} D0 + 0Y \\ E0 + FY \end{bmatrix} \\ &= \begin{bmatrix} 0 \\ FY \end{bmatrix} \\ &= \begin{bmatrix} 0 \\ \lambda Y \end{bmatrix} \\ &= \lambda \begin{bmatrix} 0 \\ Y \end{bmatrix} \\ &= \lambda V \end{aligned}$$

Thus the matrix C of eigencolumns has the same block-triangular structure as M.

Similarly, the vector $V = [\, X \quad 0 \,]$, where X is a row eigenvector of D, will be a row eigenvector of M with the same eigenvalue.

$$\begin{aligned} VM &= [\, X \quad 0 \,] \begin{bmatrix} D & 0 \\ E & F \end{bmatrix} \\ &= [\, XD + 0E \quad X0 + 0F \,] \\ &= [\, \lambda X \quad 0 \,] \\ &= \lambda [\, X \quad 0 \,] \\ &= \lambda V \end{aligned}$$

Thus the matrix R of eigenrows also has the same block triangular structure as the original matrix M.

4.5 The Effect of Non-zero Off-diagonal Elements

So far we have assumed that the matrix L is completely diagonal, with no off-diagonal non-zero elements. This is not always true.

4.5.1 Jordan blocks

When the matrix has non-zero off-diagonal entries, we can localise the problem by permuting the matrix in such a way as to bring those entries as close as possible to the diagonal. Two diagonal entries are **coupled** if there is a non-zero where their row and column intersect. Each group of coupled entries is called a **Jordan block**.

Suppose that there is a single off-diagonal non-zero element. Then we can use permutation to transform the matrix L to the structure $\begin{bmatrix} d & & \\ e & f & \\ 0 & 0 & D \end{bmatrix}$ where d, e and f are all 1×1, and D is diagonal. Then, by using the arguments on block structure, we can focus in on the top 2×2.

It is easily confirmed that provided that $d \neq f$, we can factorise $\begin{bmatrix} d & \\ e & f \end{bmatrix}$ into

$$
\begin{aligned}
L &= \begin{bmatrix} d & \\ e & f \end{bmatrix} \\
&= \begin{bmatrix} d-f & \\ e & 1 \end{bmatrix} \begin{bmatrix} d & \\ & f \end{bmatrix} \begin{bmatrix} 1 & \\ -e & d-f \end{bmatrix} /(d-f) \\
&= C'L'R'
\end{aligned}
$$

The nonzero denominator $(d - f)$ can be taken in to either C' or R'. Now we can combine C' with the previous C in $M = CLR$ and R' with the previous R, thus giving $M = [CC']L'[R'R]$, a factorisation without off-diagonals in L'.

Thus where off-diagonals appear, they link equal values on the diagonal. In general such a block can be larger than 2×2, but all the diagonal values will be equal.

4.5.2 Effect of a Jordan block on the multiplication of a (generalised) eigenvector by the matrix.

We can focus on the 2×2 case, because eigenvectors which are not associated with the block have a zero inner product with the block's rows.

Let the 2×2 matrix be $M = [\, C_1 \quad C_2 \,] \begin{bmatrix} d & \\ e & d \end{bmatrix} \begin{bmatrix} R_1 \\ R_2 \end{bmatrix}$

$$\text{Then } MC_2 = C_1 dR_1 \cdot C_2 + C_2(eR_1 \cdot C_2 + dR_2 \cdot C_2)$$
$$= C_2 d \quad \text{in the expected way,}$$
$$\text{but } MC_1 = C_1 dR_1 \cdot C_1 + C_2(eR_1 \cdot C_1 + dR_2 \cdot C_1)$$
$$= C_1 dR_1 \cdot C_1 + C_2 eR_1 \cdot C_1$$
$$= C_1 d + C_2 e$$

which is of the form $\quad \lambda(C_1 + \mu C_2)$

Thus C_1 is not an eigenvector in the sense that multiplication by M merely scales it. Every time C_1 is multiplied by M it gets a fraction e of C_2 added to it. It gets sheared within the subspace spanned by C_1 and C_2. Clearly any vector originally a linear combination of C_1 and C_2 also remains in this subspace, and is sheared in a similar way. We call the subspace an **invariant subspace** because any vector originally within it remains within it when multiplied by M.

Equally, R_2 is not an eigenvector in the strict sense, but R_1 is.

Note that the pair C_1, R_2 is not uniquely defined, because if

$$C = (C_1 + \alpha C_2)$$
$$\text{then} \quad MC = M[C_1 + \alpha C_2]$$
$$= MC_1 + \alpha MC_2$$
$$= \lambda C_1 + \lambda \mu C_2 + \alpha \lambda C_2$$
$$= \lambda(C_1 + \alpha C_2) + \lambda \mu C_2$$
$$= \lambda(C + \mu C_2)$$

and there is a similar result for R_2. Note, however, that we do still require that $RC = I$.

Effect of a Jordan block on the repeated multiplication of a (generalised) eigenvector by the matrix.

Consider the matrix $M = \begin{bmatrix} d & \\ de & d \end{bmatrix} = d \begin{bmatrix} 1 & \\ e & 1 \end{bmatrix}$.

When we raise this to a high power, n, we get

$$M^n = d^n \begin{bmatrix} 1 & \\ ne & 1 \end{bmatrix}$$

Thus the shearing effect increases arithmetically, while the overall scaling effect applies exponentially.

4.5.3 Complex eigenvalues

The set of roots of a polynomial with real coefficients can include conjugate pairs of complex numbers. Thus eigenvalues can be complex, appearing in conjugate pairs. When this happens the corresponding column eigenvectors also form a conjugate pair, as do the rows.

If we wish to remain in the real domain, this can be done by observing that the product

$$
[\,A+iB \quad A-iB\,]\begin{bmatrix} a+ib & \\ & a-ib \end{bmatrix}\begin{bmatrix} S+iT \\ S-iT \end{bmatrix}
$$
$$
= 2(a(AS-BT) - b(BS+AT))
$$
$$
= [\,A-B \quad A+B\,]\begin{bmatrix} a & b \\ -b & a \end{bmatrix}\begin{bmatrix} S+T \\ S-T \end{bmatrix}.
$$

Although the first form has fewer off-diagonal non-zeroes, the second form has only real elements, in both eigenvalues and eigenvectors. It also leads to some intuitive understanding, that if we multiply both eigencolumns by M the result is a pair of vectors rotated in the space spanned by the two eigencolumns, as well as scaled.

So again we have an invariant subspace.

4.6 Summary

Eigenfactorisation of a matrix helps us to understand what happens when vectors are multiplied by the matrix repeatedly.

5. Enclosures

5.1 Definition

An **enclosure** is a point set which contains all of the points of a given set: in our case, all the points of the limit curve or of some specific piece of it.

This allows us to test cheaply whether there are any points of the piece of curve within some test region. For example, if the enclosures of two pieces of curve have no points in common, then the two curves cannot intersect.

To be useful in this rôle, an enclosure needs three properties

(i) It must be cheap to determine from the available data, in our case the control points of the curve or a piece of it.

(ii) It must be cheap to determine whether a test point lies in it.

(iii) It must be as small as possible, so that false positives – the result that there might be a point of the curve at a particular place within it when in fact there isn't – are minimized.

5.2 Examples of Enclosures

These three are, of course, in conflict, and there are a number of possible trade-offs between them. Condition (ii) effectively demands that the enclosure should be a convex point-set, and combined with condition (iii) this leads to the use of the **convex hull**, which is defined to be the convex point set of smallest volume which contains the original set.

In this example the point set is the dark curve, and the enclosure is the shaded region.

Unfortunately, although the convex hull can be computed reasonably efficiently for 2-dimensional configurations, it becomes extremely complex for 3-dimensional ones.

We trade away from this in two ways.

(1)   We limit ourselves to plane-faced enclosures. This also helps to satisfy condition (ii). In 3D the convex hull of a finite set of discrete points is plane-faced, whereas the convex hull of a curve can have much more complex shapes. Thus the challenge becomes finding, from the representation of a curve, a finite set of planar half-spaces whose intersection is guaranteed to contain the true convex hull of the curve and thence the curve itself. One way of doing this is to use the convex hull of control points defining the curve.

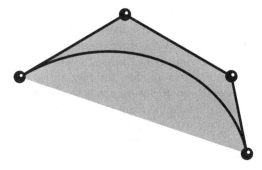

This may still be expensive in 3D.

(2)   We limit the number of orientations of those plane faces, and we choose the orientations used to be a fixed set. This makes comparison of a test point with the enclosure really fast and simple, and also the comparison of two enclosures for overlap. The best known of these, because it is the simplest, is the **bounding box**, which uses as the set of orientations those of the coordinate planes.

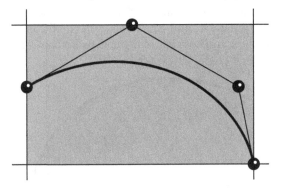

The trade-off here is that the fewer[5] the number of orientations the cheaper the setting up and the interrogation become, but the slacker the test is, giving more potential false positives.

If a tighter enclosure than the bounding box is required, then more orientations can be used.

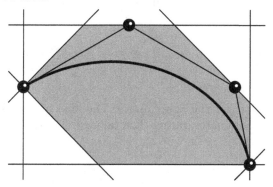

5.3 Representations

If the full convex hull is used, then it is necessary for the representation to hold both the face normal directions and the (signed) support distances from the origin to the faces.

If a fixed set of orientations is used, then the normals can be implicit, and the support distances can be held in a simple array.

Let the set of points whose enclosure is to be determined be $P_i, \subset \in 1..m$ and the set of normal directions N_j, $j \in 1..n$, and the support distances h_j. Then setting up the support distances is done by

$$\forall j, \ h_j = max_i(N_j, P_i)$$

A point Q is outside the enclosure if for any j

$$N_j.Q > h_j.$$

If the set of orientations is chosen so that

$$N_j = -N_{n+1-j},$$

then checking if two enclosures represented by support height vectors g and h overlap is just testing

$$max_j(g_j + h_{n+1-j}) > 0.$$

---

[5] Pedantically it would be possible to use just four orientations, of the faces of some fixed regular tetrahedron, but the six (plus and minus X, Y and Z) of the bounding box is a little simpler to determine and test.

Note that the normal vectors do not need to be unit vectors. Scaling does not alter the result of a sign check. Thus the set of 2D vectors

$$
\begin{array}{rr}
1 & 0 \\
1 & 1 \\
0 & 1 \\
-1 & 1 \\
1 & -1 \\
0 & -1 \\
-1 & -1 \\
-1 & 0
\end{array}
$$

would be appropriate for the figure above. The inner products can be built by just summing coordinates, rather than taking inner products with normal vectors.

5.4 Summary

(i)   An enclosure is a region of space containing some point set of interest. It is used to check quickly whether or not the point set needs to be taken into account in some computation.

(ii)   There are various widely used enclosures with different trade-offs between simplicity, cheapness of set-up and enquiries, and minimality. The minimal convex enclosure is the convex hull, but the bounding box is much simpler and is therefore most often used.

6. Hölder Continuity

One of the first questions asked by a mathematician about a function which is defined in some complicated way, which makes it clear that there isn't a nice simple closed form expression, is the level of continuity.

The first form of this question is *'Is the function continuous ?'* and if the answer is 'yes', it is rapidly followed up by *'How many continuous derivatives does it have ?'*. In fact for really interesting definitions it is possible to ask also, *'Just how continuous is the highest continuous derivative ?'*, and that is expressed in terms of the Hölder continuity exponent.

6.1 Continuity

A function $g(x)$ is said to be continuous at x if

$$\lim_{\delta x \to 0} g(x + \delta x) - g(x) = 0$$

i.e. if for any $\epsilon > 0$ we can always choose a ϕ small enough that for all $|\delta x| < \phi$, $|g(x + \delta x) - g(x)| < \epsilon$.

We have to be very careful about this definition because we are dealing with fractals. Nasty things like the Dedekind function (0 at all rationals, 1 at all irrationals) can easily slip through a less pedantic definition.

6.2 Derivatives

The derivative of $g(x)$ at x is defined to be

$$\lim_{\delta x \to 0_+} \frac{g(x + \delta x) - g(x)}{\delta x}$$

Strictly this is the **right derivative**. We can also define a **left derivative** as

$$\lim_{\delta x \to 0_+} \frac{g(x) - g(x - \delta x)}{\delta x}$$

M. Sabin, *Analysis and Design of Univariate Subdivision Schemes*, Geometry and Computing 6,
DOI 10.1007/978-3-642-13648-1_6, © Springer-Verlag Berlin Heidelberg 2010

There are associated concepts of left-continuity and right-continuity.

The value of the derivative exists at x provided that we can choose some finite value c such that for all δx small enough

$$|g(x + \delta x) - g(x)| < c|\delta x|$$

Note the difference between this and the definition of continuity, above. It is quite possible for a continuous function not to have a derivative. $f(x) = x^{1/3}$ is a good example.

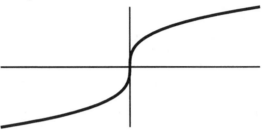

The first differences on both left and right of $x = 0$ tend to zero, but the first divided differences on both sides diverge.

The second derivative is just the derivative of the derivative, and the nth derivative is defined by recursion on n.

Continuity is defined above at a specific point (x), but we can extend this to intervals along a curve or function and to the complete curve or function by just saying that if the nth derivative of a function $g(x)$ exists everywhere in an interval (or everywhere) and it is continuous, then g is said to be C^n in that interval (or everywhere).

6.3 Hölder Continuity

It is possible to be a little more precise, telling us not just how many derivatives are continuous, but also how continuous the highest one is, on the basis of how rapidly the limit converges to zero.

Because taking the derivative drops the degree of a polynomial (or each term of a Taylor series) by 1, we do this by looking at the behaviour of fractional powers.

Consider the function

$$h(x) = \begin{cases} 0, x \le 0 \\ x^k, x > 0 \end{cases}$$

where k, which is a positive real number, has an integer part i and a fractional part f.

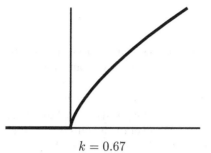

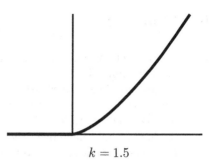

$k = 0.67$ $k = 1.5$

If $k > 0$ then at $x = 0$

$$h(x + \delta x) - h(x) = h(\delta x) - h(0)$$
$$= \delta x^k - 0$$
$$= \delta x^k$$
$$\lim_{\delta x \to 0} h(x + \delta x) - h(x) = \lim_{\delta x \to 0} \delta x^k$$
$$= 0$$

and so $h(x)$ is continuous at $x = 0$

The first derivative of $h(x)$ is

$$h'(x) = \begin{cases} 0, x < 0 \\ kx^{k-1}, x > 0 \end{cases}$$

If $k > 1$ then at $x = 0$ $h'(x)$ exists and is zero on both sides, so h' is continuous and so h is C^1.

If $0 < k < 1$ however, the function h is continuous at 0, but on the right of 0 its first derivative does not exist.

In general we can see that however high the value of k is, if it is finite then there will be some derivative which has this kind of behaviour.

The **Hölder continuity** is a pair of values, i, f. The first of these, i is integral, and is the number of derivatives which exist, the second, f, in the closed interval $(0 \ldots 1)$ which is essentially the value such that the ith derivative behaves like x^f as x approaches zero from above. These two numbers are sometimes written as $i + f$, sometimes as i, f but both indicate the same level of continuity. This leads to the notation C^{i+f} where the '+' is explicit.

A piecewise polynomial will typically have $f = 1$ at the places where the pieces meet, so that the cubic B-spline is C^{2+1} at its knots. It is, of course C^∞ over the open intervals between the knots.

The above defines the Hölder continuity at a point. The Hölder continuity of a complete curve is the lowest pointwise continuity.

6.4 Summary

(i)   The Hölder continuity of a function (and thence of a curve) is a measure of how many derivatives are continuous, and of how continuous the highest continuous derivative is.

(ii)   The notation is not totally transparent. It might have been better to have defined the Hölder *discontinuity* level as being just a real number, so that 3 would have been the discontinuity level of the cubic B-spline. C^{2+1} is sometimes evocatively written as $C^{3-\epsilon}$. Yes, the third derivative is the first to be discontinuous. But there is no point in trying to alter established conventions.

7. Matrix Norms

A **norm** is a measure of the size of an object. It is a function of some object, X, with the properties
(a) $norm(X) \geq 0$
(b) $norm(X + Y) \leq norm(X) + norm(Y)$
(c) $norm(\lambda X) = |\lambda| * norm(X),\ \forall \lambda \in \mathbb{R}$
   In this book we use norms of matrices, but the definition of these is based on the norms of vectors (finite sequences of values).
   The norm of a scalar, v, is just its absolute value $|v|$.

7.1 Vector Norms

The size of a vector, V, can be measured in a number of ways. Let the number of elements be n and the elements $V_{i \in 1..n}$.
   The important ones, called the p-norms[6], are

i   The l_1 norm, denoted by $|V|_1$, defined as

$$|V|_1 = \sum_{i \in 1...n} |V_i|$$

ii   The l_2 norm, denoted by $|V|_2$ or $||V||$, defined as

$$|V|_2 = \sqrt{\sum_{i \in 1...n} V_i^2}$$

iii   The l_∞ norm, denoted by $|V|_\infty$, defined as

$$|V|_\infty = \max_{i \in 1...n} |V_i|$$

   When we make statements which are true of any norm, then the notation $|V|$ can be used. It is no coincidence that this notation is also used for the absolute magnitude of a scalar, because that satisfies the axioms for norms.

---

[6] So-called because the generic definition is $|V|_p = \sqrt[p]{\sum_{i=1}^{n} |V_i|^p}$.

M. Sabin, *Analysis and Design of Univariate Subdivision Schemes*, Geometry and Computing 6,   33
DOI 10.1007/978-3-642-13648-1_7, © Springer-Verlag Berlin Heidelberg 2010

7.2 Matrix Norms

Each of the above vector norms induces a corresponding norm on matrices, through the definition

$$norm(A) = \max_{V \in \mathbb{R}^n \backslash 0} \left(\frac{|AV|}{|V|} \right)$$

which has the implication that

$$|AV| \leq |A||V|$$

with equality for some vector V which we can call the **support vector**.

It is possible to deduce from this definition the three properties (a),(b) and (c) above, and also a fourth property, (d), that $|AB| \leq |A||B|$.

7.2.1 Proofs of matrix norm properties

(a) $|A|$ is the ratio of a non-negative to a positive quantity which must be non-negative.

(b)   There exists a vector V such that

$$\begin{aligned}
|A + B| &= |[A + B]V|/|V| \\
&= |AV + BV|/|V| \\
&\leq (|AV| + |BV|)/|V| \\
&= |A| + |B|
\end{aligned}$$

Equality occurs when A and B share a support vector.

(c)

$$\begin{aligned}
|\lambda A| &= \max_{V \in \mathbb{R}^n \backslash 0} \left(\frac{|\lambda AV|}{|V|} \right) \\
&= \max_{V \in \mathbb{R}^n \backslash 0} \left(|\lambda| \frac{|AV|}{|V|} \right) \\
&= \max_{V \in \mathbb{R}^n \backslash 0} |\lambda| \left(\frac{|AV|}{|V|} \right) \\
&= |\lambda| \max_{V \in \mathbb{R}^n \backslash 0} \left(\frac{|AV|}{|V|} \right) \\
&= |\lambda||A|
\end{aligned}$$

(d)

$$|ABV| \leq |A||BV| \leq |A||B||V|$$
$$|AB| = \max_V(|ABV|/|V|) \leq \max_V(|A||B||V|/|V|) = |A||B|$$

It is also possible for other measures of matrix size to be defined, and they are also called norms provided that they satisfy this additional property as well as the three above.

All the above applies to matrices which are not square, but if a matrix A is square, then because, if V is an eigencolumn of the matrix, we know that $AV = \lambda V$, any eigenvalue gives a lower bound for any of the norms.

$$\left(\frac{|AV|}{|V|}\right) = \left(\frac{|\lambda V|}{|V|}\right) = \left(\frac{|\lambda||V|}{|V|}\right) = |\lambda|.$$

Clearly the magnitude of the largest eigenvalue is a tighter lower bound than smaller ones.

It is equally true that any norm is an upper bound for the largest eigenvalue.

However, the largest eigenvalue is not itself a norm because the largest eigenvalue of the product of two matrices can be larger than the product of their two largest eigenvalues. For example, the matrices $\begin{bmatrix} 3 & 3 \\ 0 & 2 \end{bmatrix}$ and $\begin{bmatrix} 2 & 0 \\ 3 & 3 \end{bmatrix}$ both have largest eigenvalue 3, while their product

$$\begin{bmatrix} 3 & 3 \\ 0 & 2 \end{bmatrix} \begin{bmatrix} 2 & 0 \\ 3 & 3 \end{bmatrix} = \begin{bmatrix} 15 & 9 \\ 6 & 6 \end{bmatrix}$$

has eigenvalues $(21 \pm \sqrt{297})/2$ of which the larger is approximately 17.1, significantly larger than 3*3.

7.2.2 Evaluating matrix norms

The definition above of a matrix norm is not directly evaluable in finite time. However, it is possible to determine the value of each of the norms from the elements of a matrix without working through all possible vectors.

We do this by choosing vectors of unit norm which can be support vectors.

In particular, $|.|_\infty$ can be evaluated by considering just n vectors, each containing a pattern of 1s and -1s matching the pattern of signs in just one row of the matrix. Thus each $|V|_\infty = 1$.

The corresponding element in AV then has a value equal to the sum of the magnitudes of the entries in that row. The largest of these determines the value of $|AV|$ and thence $|A|_\infty$. If A_{ij} denotes the ith entry in the jth row of the matrix A

$$|A|_\infty = \max_j \sum_i |A_{ij}|.$$

The l_∞ norm of a matrix is the largest value, taken over the rows of the matrix, of the sum of the absolute values of the entries in that row.

By taking a support vector with a 1 in the entry corresponding to the maximal column and zeroes elsewhere we can see that

$$|A|_1 = \max_i \sum_j |A_{ij}|.$$

The l_1 norm of a matrix is the largest value, taken over the columns of the matrix, of the sum of the absolute values of the entries in that column.

Finally, by taking as support vector the corresponding eigenvector we see that

$$|A|_2 = \sqrt{\text{largest eigenvalue of } A^T A}.$$

7.3 Summary

(i)   A norm is a way of measuring the size of the entries in a vector or matrix.

(ii)   There are several different useful norms.

(iii) For a square matrix, all matrix norms have value at least as large as the largest eigenvalue of the matrix.

8. Joint Spectral Radius

The **spectral radius** of a square matrix is the absolute value of its largest eigenvalue.

The **joint spectral radius** of two square matrices A and B of the same size, is defined by the following steps:

(i)   Let J_1 be the larger spectral radius of A and B.

(ii)   Let J_n be the nth root of the largest spectral radius of all possible product sequences consisting of n matrices each being either an A or a B, taken in any sequence. There are 2^n such sequences.

(iii)   Let R_m be the maximum value of J_n taken over all values of n between 1 and m.

(iv)   The joint spectral radius $J(A,B)$ is the limit of R_m as m tends to ∞.

We can get some handle on this value by noting that any norm is an upper bound on the spectral radius of a matrix. If, during the tending of m to ∞, it is found that the matrix given by one of the product sequences has a norm equal to its spectral radius, then that will be the joint spectral radius of the A and B.

The l_∞ norm of such a matrix is much cheaper to compute than the eigenvalues, and as m increases the nth root of the lowest norm so far of the product sequences converges to the joint spectral radius, as does (by definition) the nth root of the largest eigenvalue so far.

Other useful properties are that if A and B share an eigenvector, then both AB and BA will also share that eigenvector, and the corresponding eigenvalue of the product will just be the product of the eigenvalues of A and B. The square root cannot be larger than the larger of these factors.

In particular, the joint spectral radius of any matrix with itself is just the spectral radius of that matrix.

There is also a more general result, that if two matrices share a nested sequence of invariant subspaces, all of their products also share these subspaces, and the nth roots of the eigenvalues associated with the components within it will not exceed the eigenvalues of the originals. The proof is technical and can be found in Appendix 1 Theorem 4.

It is also true that the bounds on J_{m^2} are never looser than those on J_m, because the set of product sequences considered in J_{m^2} includes the squares of all product sequences considered in J_m. The largest eigenvalue of the square

M. Sabin, *Analysis and Design of Univariate Subdivision Schemes*, Geometry and Computing 6,   37
DOI 10.1007/978-3-642-13648-1_8, © Springer-Verlag Berlin Heidelberg 2010

of a matrix is just the square of the corresponding eigenvalue of the original, and the norm of the square of a matrix is never larger than the square of the norm of the original. Other products can have smaller norms or larger eigenvalues.

8.1 Summary

(i)   The joint spectral radius is an upper bound on what can happen when a vector is multiplied by a sequence of square matrices, each one being selected from some given set of matrices of the same size.

(ii)   For our purposes the number of matrices in that set is two, and that is interesting enough.

(iii) The calculation of the joint spectral radius from the matrices is not at all trivial, but it can be eased if the two matrices share a set of nested invariant subspaces.

9. Radix Notation

Decimal notation for both integers and fractions is an everyday familiarity. It is, however a specific case of a more general notation in which digits are used in a way where their positions carry information as well as their typography.

In the more general, **radix-r** notation, a number is denoted by a sequence of digits[7] $d_l \ldots d_{-k}$, each digit in the range $0 \ldots r-1$ to have the meaning

$$d_l r^l + d_{l-1} r^{l-1} + \ldots + d_1 r^1 + d_0 r^0 + d_{-1} r^{-1} + \ldots + d_{-k} r^{-k}$$

There are certain useful checks for divisibility of integers, well known in radix 10. For example, divisibility by 9 is checked by recursively checking the divisibility of the sum of the digits, and divisibility by 11 by taking the sum of the digits in even places and subtracting from it the sum of those in odd places, and checking the result for divisibility by 11.

These are just special cases of relationships which work equally well in this extension of the system where $r \neq 10$.

(i)   Divisibility by $r - 1$

Because r gives a remainder of 1 when divided by $r - 1$, rd gives the same remainder as d when so divided. Because $r^2 - 1$ is algebraically divisible by $r - 1$, so do $r^2 d$ and indeed any terms of the form $r^n d$. Thus the expression $d_l r^l + d_{l-1} r^{l-1} + \ldots + d_1 r^1 + d_0 r^0$ gives an initial remainder of $\Sigma_0^l d_l$. If this is itself divisible by $r - 1$, then the original expression also was.

Thus to test if a decimal integer is divisible by 9, add the digits and then test to see whether the sum of the digits is divisible by 9. You can do this recursively, and the process will always terminate because the sum of the digits is less than the original number whenever the number of digits is more than one. When there is only one digit left, it is divisible by 9 if it is equal to 9. It is also divisible by 9 if it is equal to zero, but this will not occur in a sum of the digits of a non-zero number.

If $r - 1$ is composite, then we can test for divisibility by one of its factors by initially multiplying the candidate by the cofactor and then testing for divisibility by $r - 1$. The remainder when the recursion finishes will either be

---

[7] With a 'point' between d_0 and d_{-1}.

M. Sabin, *Analysis and Design of Univariate Subdivision Schemes*, Geometry and Computing 6,   39
DOI 10.1007/978-3-642-13648-1_9, © Springer-Verlag Berlin Heidelberg 2010

$r - 1$ or something else. If it is $r - 1$, then the original number is divisible by the cofactor. This leads to the fact that the sum of digits of a number divisible by 3 is itself divisible by 3.

(ii)   Divisibility by r

The remainder on dividing $d_l r^l + d_{l-1} r^{l-1} + \ldots + d_1 r^1 + d_0 r^0$ by r is just d_0, and so $d_l r^l + d_{l-1} r^{l-1} + \ldots + d_1 r^1 + d_0 r^0$ is divisible by r iff $d_0 = 0$

A decimal integer is divisible by 10 if its last digit is zero.

A similar argument about factors gives simple rules for divisibility by 2 and 5.

(iii)   Divisibility by $r + 1$

dr has a remainder of $-d$ when divided by $r + 1$; dr^2 a remainder of $+d$, and so on alternately, so that dr^n gives a remainder of $-d$ when n is odd and of $+d$ when n is even. Thus the total remainder when $d_l r^l + d_{l-1} r^{l-1} + \ldots + d_1 r^1 + d_0 r^0$ is divided by $1 + r$ is given by alternately adding and subtracting the digits.

A decimal integer is divisible by 11 if the difference between the sum of the odd digits and the sum of the even ones is divisible by 11. Again, the test can be applied recursively to handle easily numbers of any size, terminating when the difference is 10 or less.

Because 11 is prime there are no simple examples of divisibility by its factors. However, we can regard this as an example of divisibility by a factor of $r^2 - 1$, in this case 99. This involves taking the digits in pairs, as 'digits' with respect to the radix $r^2 = 100$.

(iv)   Divisibility by $r^2 + r + 1$

We can take this further by noting that dr^3 has a remainder of d when divided by $r^2 + r + 1$. Thus $d_l r^l + d_{l-1} r^{l-1} + \ldots + d_1 r^1 + d_0 r^0$ is divisible by $r^2 + r + 1$ iff each of the three sums of digits taken by splitting the original number into consecutive triples is the same.

In the case of $r = 10$, this gives a simple test for divisibility by 37. First multiply the candidate number by 3 and then test for divisibility by 111.

9.1 Summary

Because of certain properties of divisibility of $r^2 - 1$, numbers expressed in radix notation can be tested for divisibility rather easily. The relevance here is that the radix might be algebraic, rather than a fixed number, and all of these results still apply.

10. z-transforms

The z-**transform** was originally devised by control engineers of the early 1950's who needed to extend the Laplace transform methods, which served well in determining the stability of analogue control systems, to the new world of digital control systems in which time delays played a more important role.

In our context, these transforms are approachable much more simply as a notational device which, by exploiting the correspondence between convolution of sequences and multiplication of polynomials, makes many results easier to express and to follow.

10.1 The z-transform

The basic idea is that, given two sequences, $A \equiv [\ldots a_j \ldots], j \in \mathbb{Z}$ and $B \equiv [\ldots b_k \ldots], k \in \mathbb{Z}$ and also two polynomials $A(z) = \Sigma_j a_j z^j$ and $B(z) = \Sigma_k a_k z^k$, the convolution sequence C, given by $c_l = \Sigma_j a_j b_{l-j}$ has exactly the same entries as the coefficients c_l of the product polynomial $C(z) = A(z)B(z)$.

Armed with this fact, it becomes trivial to see
(i)   that convolution is commutative and associative.

$$AB = BA$$
$$A(BC) = (AB)C$$

(ii)   that a symmetric sequence can be expressed as a convolution of shorter such sequences if and only if the corresponding polynomial can be factorised.
(iii)   that such a factorisation is unique[8].

The triviality of the equivalence applies only to finite sequences. We take on trust that the results apply also to products where one or both factors extend indefinitely in both directions. We then need to use the concept of

---

[8] Not quite. The individual polynomial factors can be multiplied by arbitrary scalar factors provided that the product of all those factors is equal to 1. We shall resolve this later by using only polynomials whose coefficients sum to 1.

Laurent Polynomial where the exponent in a power of z can be negative as well as positive or zero.

The Laurent polynomial corresponding to a given sequence is called its **Generating Function**, its **z-transform**, or its **symbol**.

In fact we can use generating functions which are not polynomial or Laurent polynomial, provided that they have a formal Taylor expansion about $z = 0$. We can then think of them as shorter notations for that expansion, and this can be extremely useful for denoting concisely sequences with an unbounded number of non-zero entries.

10.2 Why z?

This convention, of using the letter z to denote the variable, probably stems from the control systems origin of the technique. The control engineers needed a symbol which was different from s (the similarly universal symbol for the Laplace variable), but not too different. t was not possible, because it already denoted time in that context, and z just emerged as the standard choice.

Or, maybe, that $z = x + iy$ is the standard symbol for a complex variable in a domain context, and the convention goes a lot further back.

In fact a theory with much the same content, called *generating functions*, was indeed in use much earlier in combinatorics, but it was typically applied using the letter x as the argument, rather than z.

10.3 What Sort of Object is z?

When one first looks at a polynomial the expectation is that it is a function, a map from a domain to a range, and defining the domain is an important part of the semantics of the function.

This is not necessarily the case here.

It is possible to use the idea of evaluating the polynomial at a real or complex value as an aid to proving that one polynomial is a factor of another, by showing that all the roots of the first are also roots of the second. In fact in the Laplace transform the domain is definitely that of complex numbers, and [CDM91] uses this interpretation very fluently and to good effect.

However, there are four other interpretations:-

1   A purely algebraic symbol

This interpretation says that the only reason we are playing with these polynomials is to determine coefficients of other polynomials. Formal manipulations, in which the nature of the variable plays no part, and so we do not need to define anything about it, beyond the property that it can be raised to positive or negative powers.

This is strictly correct in this context, but anybody who trumpets it too loud had better find other ways of proving divisibility of one polynomial by another than that just mentioned.

2   A very large radix

This is actually a nice concretisation of the previous viewpoint. It says that polynomial multiplication and convolution are just like long multiplication, learned in primary school. However, long multiplication has a complication called *carrying* which these other two operations don't have. So we need to make the radix large enough that carrying doesn't happen. But because carrying never happens we don't have to specify exactly what value the radix has.

This viewpoint does help to make the Laurent Polynomial idea much less outlandish. You can do long multiplication with decimal fractions just as well as with integers, and the actual manipulation of the coefficients is more or less independent of where you put the decimal point.

In fact a very small radix ($<<< 1$) also avoids carrying, and this has the advantages that

(i)   the natural sequence of the entries is the same as the natural sequence of digits in a z-mal number.

(ii)   the formal Taylor expansions mentioned above become much more plausible.

3   A shift operator

Yes, of course. Multiplying a sequence by z shifts all its terms along one. Note that a term which consists of only a scalar really means that scalar times z^0. In particular the term '1' is the identity operator (don't shift it at all) rather than just a number.

4   A circulant matrix

Yet another operation which also has the same rules as convolution, multiplication of polynomials and long multiplication with a large radix, is the multiplication of circulant matrices.

z can then be interpreted as an infinite circulant matrix containing all zeroes except for the entries just below the diagonal, which are all unit values. (So that multiplying a vector by it shifts it down a row.)

Each sequence is then expressed as an infinite circulant matrix whose columns (or the reverses of the rows) are all copies of the sequence.

This analogue is most relevant when one or both of the sequences is potentially infinite in length.

Each of these ideas has its part to play in making the manipulations that we shall be carrying out in the main part of this book seem perfectly natural and obvious. They all describe exactly the same mathematics and are all, in that sense, exactly equivalent. There is no question of any one of them being 'better' or 'more accurate' than any of the others in any absolute sense, though some may be more effective than others in making obvious some specific results.

10.4 Some Special Sequences

There are some rather special sequences that we shall encounter later in the book, which have rather simple expressions in this notation. Note that just as the symbol z really means something rather abstract, the symbol '1' also means an identity operator, rather than a numerical value.

1) $(1+z)/2$

This is the sequence $[\ldots, 0, 1/2, 1/2, 0, \ldots]$. Convolution of a sequence $A : [\ldots a_j \ldots]$ with this gives the sequence of means of adjacent entries. If you want to position those means half-way in between the original entries in some sense, then the notation $(1+z)/2\sqrt{z}$ is available, but usually we don't worry too much about positioning. (Except when we are doing term by term addition or subtraction of sequences, when it has to be exactly right. Then we can use the symbol

$$\sigma = \frac{1+z}{2\sqrt{z}}$$

to denote this operation.)

2) $(1-z)$

This is the sequence $[\ldots, 0, -1, 1, 0, \ldots]$. Convolving another sequence with it gives the sequence of first differences. Again, a $\sqrt{z}$ can position the sequence in the right place for central differences if aligning sequences is really important. Again, we can give this operator a name:

$$\delta = \frac{1-z}{\sqrt{z}}$$

.

Powers of δ give higher differences.

3) $z^{-\infty}/(1-z)$

This starts to show the power of this notation. It gives the sequence $[\ldots, 1, 1, 1, \ldots]$. Not surprisingly, the product (or convolution) with $(1-z)$ gives the zero sequence ($[\ldots, 0, 0, 0, \ldots]$)(except at $-\infty$, which is well out of the way).

Taking higher powers of $1/(1-z)$ gives sequences which vary linearly, quadratically $\ldots$, and this is useful in considering the precision set of a subdivision scheme.

If you don't like the idea of using this shorthand for a polynomial with an unbounded number of terms, *which may not even converge for all values of z*, you can think of z as being a very tiny radix, which is just as effective as a large radix for avoiding carries.

Note that this implies that a sequence P whose terms P_j are j^d, can be denoted by $z^{-\infty}/(1-z)^{d+1}$. We then see easily that

$$(1-z)^{d+1}P = 0$$

(except at $-\infty$) so that the $d+1^{\text{th}}$ differences (and all higher differences) of P are zero.

10.5 Normalisation

The astute reader may well ask *"Why is $(1+z)$ divided by 2, while $(1-z)$ is not ?"*.

This is just a convention which has been found to work. The concept underneath is that we are usually operating on sequences of points. The mean of two points is another point; the difference of two points is a displacement vector, but the sum of two points does not have a clear geometric meaning.

It is a particular case of a broader convention, that we express polynomials whose coefficients do not sum to zero (or infinity) as a polynomial whose coefficients sum to one, times a scalar factor. This (or some equivalent) is necessary in order to make factorisation of polynomials fully determinate.

For example, $4z^2 + 10z + 4 = (4z + 2)(z + 2) = (2z + 1)(2z + 4)$. Which of the two factorisations is most useful ? This problem is neatly sidestepped by expressing the factorisation as

$$18 \left(\frac{4z^2 + 10z + 4}{18} \right) = 18 \left(\frac{2z + 1}{3} \right) \left(\frac{z + 2}{3} \right)$$

Taking first differences, on the other hand, is a standard operation, and dividing the first difference by two would be an arbitrary deviation from standard practice.

This convention is in fact rather nicely self-consistent. Consider the central *divided* difference sequence $(1 - z^2)P/2z$ obtained by subtracting, from each member of P the member two earlier, and dividing by the distance, 2, between them. This factorises into $\delta\sigma P$, which is exactly the sequence of first divided differences of means.

10.6 Summary

We want a notation in which to do algebra involving lots of convolutions. That notation has to make the algebra short and transparent, preferably without losing too much rigour. This is exactly what z-transforms provide.

Part II. Dramatis Personae

A univariate subdivision scheme is a set of rules by which a denser polygon is defined in terms of a sparser one. The same set of rules can then be applied again to make an even denser one, and this can be repeated indefinitely to make such a dense polygon that it looks like a curve. In principle an infinite number of such refinements would indeed give a continuous curve, and it is possible to deduce from the rules some properties of that curve without actually taking an infinite number of steps.

It is appropriate at this point to define some terms relatively precisely, which will be used in the remainder of this book.

We also introduce some particular subdivision schemes of either technical or historic interest, which will be used as examples throughout.

11. An introduction to some regularly-appearing characters

This section defines terms. Where a word or phrase appears in bold font, it is being defined by the context in which it appears.

11.1 Polygons

We start with the **original** or **initial polygon**. This consists of a sequence of **vertices**, joined, at least for display purposes, by straight line segments called **edges**. The sequence can either have two ends, or else can be cyclic, with the end leading back into the beginning. The number of and positions of the vertices define the polygon completely and we don't ask how they were chosen. This is an input defined by whoever wants to design the shape of the final curve. We therefore call the vertices **control points**. The term **point** on its own indicates a position; the term **vertex** a rôle within a polygon.

One application of the rules (a **step**) leads to the construction of a **refined polygon**. When we are talking about later steps the input to that step is called the **old polygon** and the output from it the **new polygon**.

The description above is extremely general, and initially we look at a subset about which it is possible to prove things. We consider the context in which the rules take the form of using linear combinations of the coordinates of the old polygon vertices to give the coordinates of the vertices of the new polygon at each step. The set of weights in the linear combination giving a new vertex is called a **stencil**. We take the **uniform, stationary, symmetric** case.

Uniform means that the same stencils are used everywhere along the polygon. A **non-uniform** scheme could have stencils near one end of the polygon different from those near the other and yet again different everywhere in the interior.

Stationary means that the same stencils are used at every step of refinement. A **non-stationary** scheme could have different stencils used for each step.

Symmetric means that the shape of the limit curve does not depend on which end of the polygon is regarded as the start and which one as the finish.

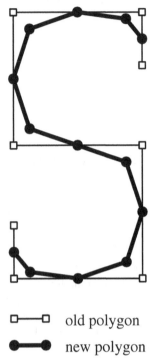

old polygon

new polygon

Example Scheme 1

For example, in one particular scheme each step creates alternate new vertices (**e-vert-ices**) at the midpoints of the edges of the old polygon (i.e. at a weighted mean with weights $1/2, 1/2$) and intervening ones (**v-vertices**) at points given by a weighted mean of three consecutive vertices with weights $1/8, 6/8$ and $1/8$ respectively. Both from the symmetry of these weights and from the fact that the dominant weight is the central one, each new v-vertex is associated with the central old one of the three used to define it.

These new vertices are joined together, in the sequence matching the sequence of the edges and vertices of the old polygon, to form the new polygon.

This particular scheme is called the cubic B-spline scheme, because its limit curve is indeed a cubic B-spline curve.

At the ends it can be convenient for a new v-vertex, which does not have three old vertices to apply the weights to, merely to be placed at the same position as the end vertex of the old polygon. Much later in the book we shall see why this is an unfortunate end-condition, but for now it provides a simple way of tying up the loose end.

When we are dealing with just a local part of a polygon somewhere in the interior, so that end-conditions do not complicate the story, we will refer to **'local configuration'** or **'configuration'** rather than 'polygon'.

This first example has approximately twice as many vertices in the new polygon as in the old. We call it a **binary** scheme. If there had been three times as many it would have been a **ternary** scheme, and such generalisations will be discussed in a few pages' time. In principle at each refinement we can multiply the number of vertices by whatever we choose, and this number is called the **arity** and denoted by the letter a. It is also called the **dilation factor**, which stems from generating function usage.

11.2 Labelling and Parametrisation

When we start to apply some mathematics to these objects, we need a way of identifying how the correspondence between old and new vertices works. The method used here is that the old vertices are labelled by integers, multiples of the arity[9]. Their positions will be denoted by P_i, $i \in a\mathbb{Z}$ when we get to some mathematics. The v-vertices get the same labels as the old vertices with which they are associated, and the e-vertices after one step get the intervening integers. These positions will be called p_i, $i \in \mathbb{Z}$. In the binary case the v-vertices get even labels, the e-vertices odd labels. Applying upper case for old vertices and lower case for new allows us to use the same letter for two different levels of refinement without a lot of extra superscripts or subscripts.

When subsequent steps are applied, new e-vertices get first half-integer labels, then quarter-integer etc. and so successive steps fill in all the **dyadic numbers**[10].

These are dense in the reals and so in the limit we have something very close to a continuous parametrisation of the limit curve using vertices alone. However, we can extend the labelling to a continuous parametrisation at every stage by associating (by linear interpolation) intermediate labels with the points on the edges of the polygon.

Every polygon is thus a parametric curve, and so is their limit, the limit curve. To every real value of parameter (and we shall use the letter t to denote the parameter) between 0 and the arity times the number of original sides of the polygon there corresponds a point of the limit curve[11]. It may not be possible to write a closed form for this function (except in some special cases) and the function may not be differentiable or even continuous, but it is in principle defined at every real value in the domain, as the limit of a sequence of points lying somewhere on consecutive polygons[12].

---

[9] In fact we can put the origin, the label of value zero, at any convenient point, because all of the mathematics is invariant under a consistent translation of the whole configuration.

[10]The dyadic fractions are those which have a finite representation as binary numbers. Rationals which are not dyadic have a binary fraction representation which after a while repeats some pattern to infinity. Irrationals do not have any such pattern in their infinite binary representation

[11]This is not quite true, because some schemes have special rules at the ends which allow for definition of new vertices which don't have old vertices on both sides. Others do not, and then each new polygon covers a slightly shorter parametric range than the previous one. This distinction can be ignored until we come to the chapter on end-conditions at page 175.

[12]There will be a somewhat more sophisticated viewpoint taken later (see page 115) but this is a good enough first approximation to bootstrap the ideas.

11.3 Primal and Dual Schemes

The example above is called a **primal** scheme, because old vertices map into new v-vertices under the labelling.

However, it is also possible to define schemes in which the relationships are more subtle.

Example Scheme 2

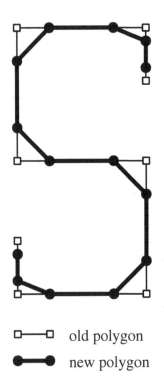

◻——◻ old polygon

●——● new polygon

In this scheme, which is called the quadratic B-spline scheme, because its limit is indeed a quadratic B-spline curve, new vertices are constructed at points one quarter and three quarters of the way along each edge of the old polygon. At the first step they get labels which are half-integers.

Such schemes are called **dual** schemes because edges map into edges under the labelling.

The terms e-vertex and v-vertex are no longer applicable, but the labelling is clear. It would be possible to use the terms 'e-edges' and 'v-edges', but in fact we don't bother.

It is tempting to say that the parameter values of new vertices are obtained by applying the same process to them as to the coordinates. However, this is dependent on a subtlety called **linear precision** which will be elaborated later in chapter 20 below.

11.4 Ternary Schemes and Higher Arities

Both of the above examples approximately double the number of vertices in the polygon with each step of refinement. They are **binary** schemes. It is also possible to have schemes in which the number of vertices trebles or quadruples or is multiplied by a still higher factor. As mentioned above, we call that factor the **arity**, so that binary schemes have an arity of 2, **ternary** of 3, **quaternary** of 4 etc. Some of the mathematics applies to all arities, and in such cases we will denote the arity by the letter a.

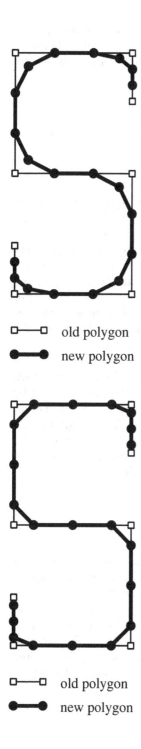

□——□ old polygon

●——● new polygon

□——□ old polygon

●——● new polygon

Example Scheme 3

One particular ternary scheme, called the **ternary quadratic B-spline** for the obvious reason, has new v-vertices given by weights of $1/9$, $7/9$, $1/9$, and new e-vertices given by the weight combinations $2/3$, $1/3$ and $1/3$, $2/3$.

Clearly the old vertices have labels which are multiples of 3. The new e-vertices get labels of integers not divisible by 3. Such a scheme is both primal and dual, because vertices map into vertices and also edges map into edges. We call it a '**both**' scheme.

This particular scheme turns out to have exactly the same limit curve as the second example above.

Example Scheme 4

It is also possible to have more complicated schemes in which vertices map under the labelling into edges and edges into vertices, so the scheme is neither primal or dual.

The terms e-vertex and v-vertex are no longer applicable, but the labelling is clear. After one step there are new vertices with fractional labels.

An example has these vertices defined by weight combinations $[5/6, 1/6]$, $[1/2, 1/2]$ and $[1/6, 5/6]$. This is a paradoxical scheme whose limit curve has strange properties that we shall deduce as examples later in the book. We call it just the 'ternary neither scheme', as its main function is to illustrate that possibility.

For precision we shall refer to schemes of odd arity which are both primal and dual as **both** schemes, and to schemes which are neither primal or dual as **neither** schemes, but the reader should be aware that other authors refer to them as 'primal' and 'dual' respectively.

11.5 Interpolatory Schemes

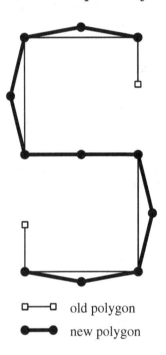

□——□ old polygon

●——● new polygon

Example Scheme 5

All the previous examples have the property that the corners are smoothed off the original polygon. It is also possible for a scheme to have the property that the new polygon has new v-vertices which lie exactly at the corresponding old polygon vertices. If this is true after one step it will also be true after two, or three, or more, and indeed it is true, by induction, of the limit curve.

The scheme illustrated here has its v-vertices at the original vertices, and each of its e-vertices at the place defined by a parametric cubic Lagrange interpolant through four points. The weights for an e-vertex are -1/16, 9/16, 9/16 and -1/16, and the scheme is therefore known as the **four-point scheme**.

11.6 Range

The limit curve of a subdivision scheme is a function from real parameter values. The range of points in the illustrations, particularly in this chapter, is two-dimensional Euclidean space, represented by $\mathbb{R}^2$. In typical CADCAM or animation usage they will be in three-dimensional Euclidean space, represented by $\mathbb{R}^3$, though it is possible for even more dimensions to be involved if texture coordinates or temperatures or values of other properties are handled in parallel with the three coordinates. This variation of range is rendered trivial by the fact that each coordinate or other property is handled independently of the others[13].

---

[13]At least in the limited set of schemes considered most of the way through the book.

We can do most of our analysis by considering just a single real ordinate as the range. When the map is to a single real, it is convenient to plot the abscissa t as the x-coordinate and the value as y. Some of the figures in later sections are plots of this kind. It should be easily visible whether a figure is one of these or, like the earlier ones in this chapter, a plot of $y(t)$ against $x(t)$, both being functions of t.

11.7 Representations of Subdivision Schemes

There are four ways in which we can look at the weighted mean coefficients which define how a subdivision scheme behaves: stencils, the subdivision matrix, the mask, and the generating function.

Stencils   The first obvious representation is the set of weight combinations for each of the types of vertex in the new polygon.

Each set of weights is called a **stencil**, and the set of stencils taken together is a complete description of a uniform, stationary scheme. (The arity is implicit in the number of stencils.)

Because the values in a stencil are often rational with a relatively small denominator, it saves space when writing to put a common denominator outside the brackets which enclose the values. Because the values are components of a weighted mean, the sum of the values in a stencil must be unity, so the denominator can always be derived from the set of values themselves.

The stencils of example scheme 1 are thus [1,6,1]/8 for v-vertices and [4,4]/8 for e-vertices. There are two different stencils because the arity is 2.

Example scheme 2 also has two stencils, [3,1]/4 and [1,3]/4 . Example scheme 3 has three stencils because it is a ternary scheme, [3,6]/9, [1,7,1]/9 and [6,3]/9.

Subdivision Matrix   It is also possible to assemble the stencils into a matrix, by which the column vector of old vertices is multiplied to give the column vector of new ones.

The stencils are clearly visible as the rows. Because the scheme is uniform, the same stencils alternate all the way down the matrix. If the scheme had been ternary the same sequence of three stencils would have repeated all the way down.

One step of the scheme of example 1 can be expressed as the matrix

$$8\begin{bmatrix} \vdots \\ \vdots \\ p_{i-2} \\ p_{i-1} \\ p_i \\ p_{i+1} \\ p_{i+2} \\ \vdots \\ \vdots \end{bmatrix} = \begin{bmatrix} \ddots & & & & & & \\ 1 & 6 & 1 & & & & \\ & 4 & 4 & & & & \\ & 1 & 6 & 1 & & & \\ & & 4 & 4 & & & \\ & & 1 & 6 & 1 & & \\ & & & 4 & 4 & & \\ & & & 1 & 6 & 1 & \\ & & & & 4 & 4 & \\ & & & & 1 & 6 & 1 \\ & & & & & & \ddots \end{bmatrix} \begin{bmatrix} \vdots \\ P_{i-4} \\ \\ P_{i-2} \\ \\ P_i \\ \\ P_{i+2} \\ \\ P_{i+4} \\ \vdots \end{bmatrix}$$

Mask   Just as the matrix is made up of the stencils as rows, we can also think of it as made up of columns. When a larger part of the matrix is drawn, rather than just a short section, it becomes visible that all the columns are the same, merely being shifted down 2 rows (or in general the same number of rows as the arity) for each step to the right.

Such a column is called the **mask** of the scheme. That of example scheme 1 is $[1, 4, 6, 4, 1]^T/8$, but when it is clear that we are talking about masks the transposition will not be written in every time.

Example scheme 2 has as its mask $[1, 3, 3, 1]^T/4$, example scheme 3 has mask $[1, 3, 6, 7, 6, 3, 1]^T/9$ and example scheme 4 has $[1, 3, 5, 5, 3, 1]^T/6$. The sum of the values in the mask is the same as the total of the sums of the values in the stencils, and so it is equal to the arity. This means that to complement the actual string of numerator values either the denominator or the arity is needed to be completely precise. If neither is quoted, the default assumption is that the scheme is binary.

Some early papers referred to the stencils as the masks, but this usage can be distinguished by the use of the plural.

The stencils show diagrammatically the influences of neighbouring old vertices on a given new one: the mask shows diagrammatically the influences of a given old vertex on neighbouring new ones.

In a symmetric scheme, the mask is palindromic: each stencil either is palindromic, or else has a mirror-image mate.

In order to make clear exactly how the mask is applied, the entries are given the same labels (parameter values) as the new vertices influenced by P_0. If the mask entries are denoted by m_t, this makes the symmetry relation particularly simple,

$$m_{-t} = m_t.$$

The new vertices are then given by

$$p_t = \sum_{s \in a\mathbb{Z}} m_{t-s} P_s$$

Note that t here might not be an integer: for binary dual schemes each t will be an odd half-integer. s will be an integral multiple of the arity. These complicated conventions are designed exactly to make these last two equations simple.

Generating Function   The fourth representation is made by treating the values in the mask as the coefficients of a Laurent Polynomial. This is the z-**transform**, which maps a sequence of values into a function. The Laurent polynomial is also called the **symbol** of the scheme.

This is not an obvious thing to do, but it turns out to be incredibly powerful. The immediate application is that convolution of the mask with the old polygon becomes multiplication of the generating function of the old polygon by the generating function of the mask.

Thus one step of example scheme 1 can be written as

$$\ldots + p_{i-2}z^{-2} + p_{i-1}z^{-1} + p_i z^0 + p_{i+1}z^1 + p_{i+2}z^2 \ldots$$
$$= (1z^{-2} + 4z^{-1} + 6z^0 + 4z^1 + 1z^2)*$$
$$(\ldots + P_{i-4}z^{-4} + P_{i-2}z^{-2} + P_i z^0 + P_{i+2}z^2 + P_{i+4}z^4 \ldots)$$

Note that because the spacing of the old vertices is twice as sparse as that of the new ones, the old Laurent polynomial is in z^2.

If the scheme is not primal, the z-transform then becomes a generalisation of a Laurent polynomial: a Laurent polynomial with a shift of a fractional power of z. This shift is not important for the algebra: the entire equation is just shifted a little to one side, but the shift in the algebra is actually achieving the maintenance of symmetry.

These four representations all convey exactly the same information. All are equivalent. We shall choose different ones for different purposes through this book. Stencils are easier for first describing a scheme; the mask is a more concise and precise representation; the matrix and the generating function are representations which lend themselves to different analyses.

All are exactly equivalent, holding the same information about the scheme. Because they are so tightly linked, we shall sometimes use somewhat sloppy wording, saying, for example, that scheme $S1$: $([1, 4, 6, 4, 1]/8)$ has a factor of $[1, 1]/2$, when what has the factor (of $(1+z)/2\sqrt{z}$) is the corresponding generating function. If the name of a scheme is $S1$, for example, we shall refer to 'the matrix $S1$' or to '$S1(z)$' without further comment. Please bear with this: it makes the reading much easier and does not cover up lack of rigour in the argument.

11.8 Exercises

(i)   For each of the schemes in 11.9(i), write out the stencils of the scheme. Use the notation which inserts a '*' at the place of an edge-vertex.

(ii)   For each of the schemes in 11.9(i), write out the part of the matrix with non-zero principal diagonal. What complications did you find in interpreting this question ?

(iii)   How can the denominator of a scheme be determined from the arity and the sequence of integers in the 11.9(i) examples ? Equally, how can the arity be determined from that sequence and the denominator ?

(iv)   Write a small program which loads into a memory data structure the arity and mask of a scheme. A convenient structure holds the arity and the number of entries as integers and the mask as an array of floating point numbers. The source of the data should be either a file or values entered through the screen by the user. This question will not have a solution provided, but questions in later chapters of the book will extend this program to carry out analyses on masks held in this data structure.

11.9 Summary

(i)   Five example schemes have been encountered.

Scheme	Arity	Name	Mask
1	2	Cubic B-spline	$[1, 4, 6, 4, 1]/8$
2	2	Quadratic B-spline	$[1, 3, 3, 1]/4$
3	3	Ternary Quadratic B-spline	$[1, 3, 6, 7, 6, 3, 1]/9$
4	3	Ternary neither	$[1, 3, 5, 5, 3, 1]/6$
5	2	Four-point	$[-1, 0, 9, 16, 9, 0, -1]/16$

Between them these illustrate many of the aspects to be analysed in later chapters.

(ii)   Subdivision curves are parametric curves. They can be maps from the reals to curves in a space of any number of dimensions. The important numbers of dimensions are

1:   This gives the *functional* form. In fact when dealing with higher dimensional spaces we deal with each of the coordinates independently.

2:   for illustrations in this book, where the pages are two-dimensional.

3:   for real applications, where twisted curves are often required.

        Clearly there is little difference between a two-dimensional curve and a curve which happens to be planar. Also, if a scheme has the property called **linear precision** (which all useful schemes do in fact have), the abscissa itself obeys the same rules as the ordinate(s), and so even the functional case is just a particular case (not a special case) of the general n-dimensional one.

(iii) There are four representations of a uniform stationary scheme.
- Stencils
- Matrix
- Mask
- Generating Function or Symbol

These are all exactly equivalent in terms of information content. The stencils appear as the rows of the matrix and the mask as the columns. The generating function is a nice notation, associating some semantics with the entries in the mask, which allows us to do some algebraic manipulations on the mask in an easily understood way. We shall use the mask as the definition of a uniform stationary scheme.

This chapter has introduced a large number of new terms, but we have now reached the point that we can embark on analysis. A colleague has just brought in a new scheme and asked *"I think this new scheme (here is the mask) is marvellous: what do you think of it ?"*, and the next few chapters will allow us to find out systematically whatever weak points that scheme has.

Part III. Analyses

The body of this book describes five major questions which we are now able to answer about an arbitrary uniform, stationary, symmetric subdivision scheme. In each case there are additional issues which are conveniently addressed in the same chapter.

- Support: How much of the limit curve is influenced by a given control point ?
- Enclosures: How closely can we put boxes around parts of the curve to help test for intersections etc. ?
- Continuity: What discontinuities of some derivative are present in the limit curve ?
  This is a long story with several aspects.

  - necessary conditions from eigenanalysis
  - sufficient conditions from z-transform analysis
  - using deeper factorisation of the z-transform
  - combining the necessary and sufficient conditions, using the Joint Spectral Radius.
- Precision Set and Order of Approximation: What degree of polynomial can be reproduced exactly, and how does the error between the limit curve and a curve from which the control points are sampled vary with the density of sampling ?
- Artifacts: What features can be seen in the limit curve which cannot be controlled by choice of the initial input polygon ?

For most purposes we look at subdivision functions rather than subdivision curves, because subdivision curves can be looked at as parametric curves where each of the coordinates of a point on the curve is an independent function of an implicit parameter. Part way through this story we shall discover that, for almost all of the schemes of interest, the functional case is a particular case of the more general one anyway.

As we look at different analyses in the following chapters we shall take the binary primal subset first, to establish ideas, and then generalise to duals and to higher arities.

12. Support

This is the most basic and simplest of the analyses that we can do. It determines *how much of the limit curve is modified when one control point is moved.* That part which is dependent on a given control point is called the **support region** of that point. Pedantically both of these refer to an interval in parameter space, but since that corresponds to a piece of the curve we can also use 'support' to refer to that.

It is of interest for three reasons:

(i)   It is directly relevant to the ease of use of a scheme in interactive graphical editing, where the limit curve is pulled and pushed by the movement on the screen of control points[14].

(ii) Its results are used by other analyses.

(iii) It leads to the concept of the **basis function**.

12.1 The Basis Function

This is such an important concept that we take it first.

The **basis function** is the limit function resulting from **cardinal data**, where all vertices of the polygon have value zero except for one. Clearly there is one such basis function for each control point in the polygon, but in **uniform** schemes, where the weights in the weighted means do not depend on position in abscissa space, all of these basis functions have the same shape. They are just translates of each other, and so there is only one shape, which we call *the* basis function.

This gives a more precise meaning to the term **support width**. It is the width in abscissa units of the closure[15] of the abscissa region over which the basis function is non-zero (the **support region**).

---

[14] As well as the strict definition of support just made, a useful measure is the width of the region in which the magnitude of the support function is greater than some positive constant, modelling the region in which a perceptible difference in the curve can be seen.

[15] 'closure' is a technicality which allows us to ignore the fact that the basis function can be zero at isolated points of the support region.

M. Sabin, *Analysis and Design of Univariate Subdivision Schemes*, Geometry and Computing 6,   63
DOI 10.1007/978-3-642-13648-1_12, © Springer-Verlag Berlin Heidelberg 2010

In all schemes which are linear (i.e. the mask does not depend on the coordinates of the control points[16]), the limit curve can be expressed as

$$P(t) = \sum P_i b_i(t)$$

where $b_i(t)$ is the basis function associated with the ith control point. In the uniform case this simplifies to

$$P(t) = \sum P_i b(t - i)$$

because all basis functions are just translates of the same function $b(.)$.

Most of the properties that we ask about in general of limit curves are directly related to properties of the basis function.

In a few (important) special cases the basis function has a closed form, but more generally adequate approximations can be evaluated numerically by the **cascade algorithm**. This amounts to just applying the subdivision process repeatedly to initial cardinal data. There is an efficient way of doing this.

The first step of applying a scheme to cardinal data produces a sequence of control values which are just the entries in the mask, two per original span. The second step produces four control values per span. These could have been produced directly by a quaternary scheme (of arity equal to 4) with the same basis function as the original binary one, and the values just computed are indeed the mask of that quaternary scheme. In the usual description of the cascade algorithm the third step would have produced eight control values per span, the fourth 16 and the fifth 32. However, if we apply the quaternary scheme to the result of the second step we get 16 control values per span, and we can regard these control values as the mask of a scheme of arity 16. Applying this scheme to the latest values gives a scheme of arity 256 in only four steps. For graphical purposes this is typically enough, but one more step gives a representation of the basis function with 65536 control values per span which would be dense enough even for engineering applications.

As described above, the cascade algorithm is regarded as producing the basis function as the pointwise limit of a sequence of polygons. There is, however, another way of looking at it.

The limit curve is

$$P(t) = \Sigma_i P_i b_i(t)$$
$$= \Sigma_i P_i b(t - i)$$

But $P(t) = \Sigma_j p_j b(2t - j)$
$$= \Sigma_j \Sigma_i P_i m_{j-2i} b(2t - j)$$
$$= \Sigma_i P_i \Sigma_j m_{j-2i} b(2t - j)$$

Therefore $\forall i, \quad b(t - i) = \Sigma_j m_{j-2i} b(2t - j)$

---

[16]This is a very weak precondition. Even if the mask depends on position along the curve and on the step of refinement, there is still a set of well-defined basis functions.

If we take the particular case $i = 0$ (all the others are just translations of it) we get
$$b(t) = \Sigma_j m_j b(2t - j)$$
and this equation holds at all values of t.

The basis function can be expressed as the sum of scaled copies of itself, and the scaling factors are just the entries in the mask. The cascade algorithm can then be viewed as making the basis function from tinier and tinier copies of itself as iteration proceeds.

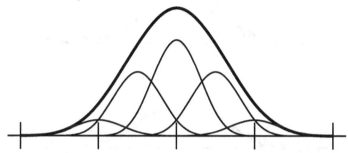

An even more important theoretical result is that if we take some other function (with some small print) $b_0(t)$ and apply the cascade algorithm by iterating
$$b_{k+1}(t) = \Sigma_j m_j b_k(2t - j)$$
successive b_k converge towards the true basis function. This property holds for all t, not just the dyadic values which we actually construct.

12.2 Support Width

We now look at the question again, of how much of the curve is influenced by a single control point.

12.2.1 Primal binary schemes

Suppose that the scheme is a primal binary one. The mask has an odd number of entries, m. Thus the effect of the original control point, P_j, being moved reaches as far after one iteration as $(m - 1)/2$ new points from P_j, or $(m - 1)/4$ old points. At the second step the mask being applied has the same numbers but is at a narrower abscissa scale, and so the second step extends the influence by one half of this, $= (m - 1)/8$, and all subsequent steps give a total series of $((m - 1)/4)(1 + 1/2 + 1/4 \ldots)$ which is easily summed to give $(m - 1)/2$.

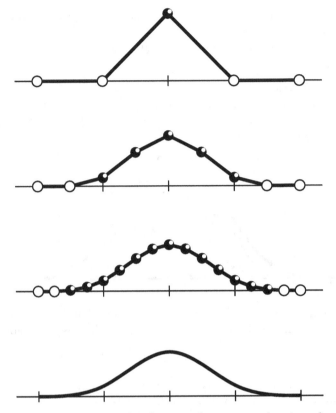

By looking at the extent of influence of one control point after 0,1,2,∞ refinements, in the cubic B-spline scheme we can see that the refined polygons converge towards the basis function, and the last non-zero entry converges towards the end of the support region.

Thus moving a single control point influences at most a piece of the limit curve which stretches $(m-1)/2$ old spans on each side. m was odd and so this is an integer.

Beyond this point the control point has no influence. We should not therefore be surprised to find a discontinuity of some derivative at such places. Note that the sum of the series converges from below, and the amount of influence actually felt at the limit of the series has diminished to zero by the time that the edge of the support is reached. The original control point can influence some derivative on the near side of the final point, but not the point itself or anything beyond it.

We have made a big assumption so far, that the subdivision process converges to a limit curve. We shall discover later that it is possible to have a scheme which is not convergent. However, these take deliberate perversity to construct, and the support analysis applies even to such monsters.

12.2.2 Dual binary schemes

Here the mask has an even number of entries. The calculations above follow through in exactly the same way, except that the range of influence of an original control point reaches to a point half-way between images of control points. This means that we expect there to be discontinuities of some derivative at these half-way points, rather than at the images of control points.

For example, the binary scheme whose mask is $[1, 3, 3, 1]/4$ (the binary quadratic B-spline) has a support width of $(4 - 1)/2 = 3/2$ spans on each side.

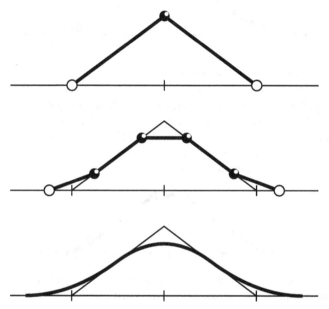

We see the same effect in the growth of the extent of influence of one control point after $0,1,\infty$ refinements in the quadratic B-spline scheme. Again, the refined polygons converge towards the basis function.

12.2.3 Ternary schemes

We consider first ternary schemes which are both primal, in the sense that there is a vertex in the new polygon corresponding to each vertex of the old, and dual, in that there is an edge of the new polygon matching each each of the old.

In this case the mask has an odd number of entries, and the distance reached at the first step is $(m-1)/6$. Considering subsequent steps multiplies this by a factor of $(1 + 1/3 + 1/9 + ...)$ which is equal to $3/2$. Thus the range of influence is $(m - 1)/4$ on each side.

Depending on the size of the mask, this can lie either at the image of an original control point or half-way in between.

For example, the ternary scheme whose mask is $[1, 2, 3, 2, 1]/3$ (the ternary linear B-spline) has a support width of exactly $(5 - 1)/2 = 2$ old spans, while the scheme whose mask is $[1, 3, 6, 7, 6, 3, 1]/9$ (the ternary quadratic B-spline) has a support width of $(7 - 1)/2 = 3$ spans, one and a half on each side of the control point.

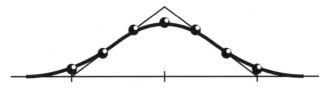

12.2.4 'Neither' ternary schemes

It is also possible for the mask of a ternary scheme to have an even number of entries. In this case exactly the same algorithm tells us that the range of an original control point can lie one quarter of the way between point images. The end of the influence of one control point is no longer at the same place as the start of the influence of another.

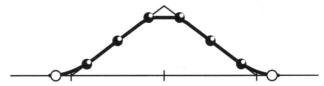

This can be seen in the basis function of the ternary scheme $[1,3,5,5,3,1]/6$ with the control points after one step from cardinal data.

12.2.5 Higher arities

As far as can be seen, all the interesting effects are covered above, and higher arities introduce no new features.

In general the width, w, of the non-zero part of the basis function is $(m - 1)/(a - 1)$ old spans, (or $(m - 1)/2(a - 1)$ on each side).

12.3 Facts which will be Relevant to Other Analyses

Once we know how much of the curve is influenced by each original control point, we can determine how many control points influence a given part of the limit curve. This is a key parameter to some of our other analyses.

Call the parameter value at which the zero and non-zero parts of the basis function meet an **end-point**.

We have to consider two cases: such a point may be either at the ends of two different control points' supports, which will be the case if the support width is integral, or at the end of only one, which will be the case if the support width is fractional.

Consider first the case of integer support width. If the support width is even, the end-point will have an integer label and correspond to an original control point. If the support width is odd, the end-point will have a half-integer label and correspond to a midedge of the original polygon.

There are three slightly different variants of the question that we need to ask.

(i)   *How many control points influence a specific point which is the end of an original control point's support ?*
Provided that the basis function is continuous[17] only $w - 1$, because there are only $w - 1$ such end-points in the non-zero part of the basis function.

In the cubic B-spline scheme, for example, the support is 4 spans wide and each end-point is influenced by only 3 original control points.

(ii) *How many control points influence the neighbourhood of an end-point ?*
Exactly $w + 1$, because there are $w + 1$ end-points in the closure of the support.

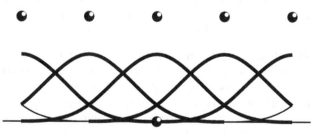

In the cubic B-spline scheme the neighbourhood of each end-point is influenced by 5 original control points.

---

[17]If the basis function is not continuous the value can be w.

(iii) *How many control points influence the interior of a specific span of the limit curve between adjacent end-points ?*

Only w, because each basis function is non-zero over the interiors of w spans.

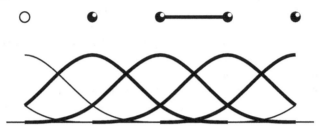

In the cubic B-spline scheme each basis function has four non-zero spans, and thence each span is influenced by 4 original control points.

Now consider the case when the support width w is not an integer: there are two kinds of spans between end-points, and they must alternate. When moving along the curve, when leaving a support region the number of active control points reduces by one, so that spans influenced by $\lceil w \rceil$ are separated by those influenced by $\lfloor w \rfloor$. The end-point itself and one side of the neighbourhood will be influenced by the lower of these figures; the other side of the neighbourhood by the higher.

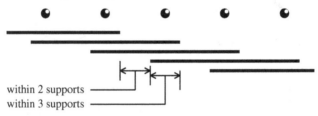

within 2 supports
within 3 supports

For example, the ternary scheme whose mask is $[1, 3, 5, 5, 3, 1]/6$ has a support width of $(6-1)/4 = 5/4$ old spans on each side, so that the part of a span near the original control points is influenced by three original vertices, while the part from $1/4$ to $3/4$ is influenced by only two.

This implies that the structure of the central part of each span can be different from the end parts. In the case just considered, the central part of each span is straight, because it is influenced by only two control points, while the part from $3/4$ along one span to $1/4$ along the next has a fractal structure. In fact the basis function has the interesting combination of properties that it is entirely made up of pieces of linear functions at different scales, but it has a continuous first derivative.

12.4 The Matrices of Powers of a Scheme

Applying a scheme to cardinal data gives its mask, because, thinking of the scheme in terms of a matrix, cardinal data essentially picks out one column from the matrix of the scheme. Applying the scheme a second time (multiplying the mask by the matrix) gives the mask of a scheme which is of higher arity. Because applying this scheme to any data is just applying the original, two steps at a time, it must have the same limit curve. We call this scheme the **square** of the original. The matrix of the square has a longer column and a steeper slope than the original. Taking yet more steps gives the masks of schemes of even higher arity and even steeper slopes.

The number of non-zero entries within any row remains constant, at the number of control points influencing one point of the limit curve. In the limit we can imagine a 'matrix' which has continuous rather than discrete columns, each of which has its value varying as the basis function. Although the 'columns' are infinitely long, the number relevant at any point is still only a relatively small number, the same as the size of the largest original stencil.

12.5 Practical Support

Our original support question was " *how much of the limit curve is modified when you move one control point ?*". When building interactive design software it may be very relevant that in regions where the basis function has a tiny value, the change in the curve from a small movement of a control point may be undetectable.

Suppose that a control point is being moved by 50 pixels interactively. In places where $b(t)<0.01$ this is unlikely to change the image of the curve. Even for the cubic B-spline the last one-third of a span at each end is well below this threshold, as is the entire end span of the quintic B-spline. For higher degrees even more will have no visible effect.

12.6 Exercises

(i)   For each of the schemes of 11.9(i) above, determine the support.

(ii) What is the square of $[1,3,3,1]/4$?, what is its arity, what are its stencils, and what is its support ?

(iii) Write a routine to take your representation of the mask of a scheme (as in question (iv) on page 58 above) and compute its square.

(iv) Using that routine, plot an approximation to the basis function of a given mask.

(v)   Write a routine to take your representation of a mask, and compute the support and also good estimates of its practical support at levels of 1%, 2% and 5%.

12.7 Summary

(i)   Only a finite part of the limit curve is influenced when one control point is moved. The length of the abscissa of that part is given (in units of old polygon edges) by

$$w = (m - 1)/(a - 1).$$

(ii)   The amount which is noticeably influenced is usually less.

(iii)   We should not be surprised to find a discontinuity of some derivative at a place where the influence of a control point ends.

(iv)   Only a finite number of control points influence each point of the limit curve.

13. Enclosure

Promenade

Knowing the support tells us how much of the limit curve is influenced when one control point is moved. We also want to know how the overall position of the curve is influenced by the set of control points as a whole. This is particularly important when calculating intersections of, for example, a subdivision curve with some plane. We express this in terms of enclosures, simply shaped pieces of space within which we can guarantee that the curve lies.

13.1 Positivity

For many schemes every point of the limit curve is a positive weighted mean of the original control points. The limit curve then lies inside the convex hull of those points.

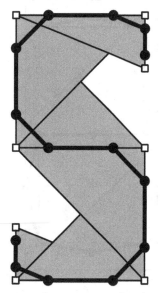

An even stronger result is that each span of the curve has its points depending on only a relatively small number of control points, and so that span lies inside the convex hull of a local group of control points, and the curve itself lies inside the union of those local hulls.

For example, convex hulls are shown for sets of three consecutive points using the scheme $[1,3,3,1]/4$. The limit curve must lie within the shaded area.

Schemes which satisfy this property are called **positive** schemes, and we first look for ways of checking whether we can rely on this.

The sharp test is that if every point on the curve can be guaranteed to be a positive linear combination of some set of control points, then we can find an enclosure for the complete curve by taking all the control points, or for any part of it, from one span up, by taking the relevant subset of the control points.

This will hold when all of the basis functions are non-negative, because it is the basis function values which give the coefficients of the linear combination of control points for a given curve point.

If any basis functions are negative at some abscissa, t, then we can set up a set of control points which force the limit point at t to lie outside the convex hull of the control points. We need only consider one coordinate, and we just give those control points for which the basis function is negative at t the coordinate $-h$, and those for which it is positive the coordinate $+h$. The linear combination will then have a value greater than h, which lies outside the convex hull.

13.2 If the Basis Functions are Somewhere Negative

The condition on a non-negative basis function is essentially a condition on the l_∞ norm of the matrix of basis functions, introduced in the last chapter, $\max_t \Sigma_i |b(t-i)| \leq 1$ which, because $\Sigma_i b(t-i) = 1$, cannot be less than 1.

Consider the argument above, which said that the convex hull only worked if l, the l_∞ norm, was less than or equal to 1. This same argument can still be applied to construct an enclosure when $l \geq 1$.

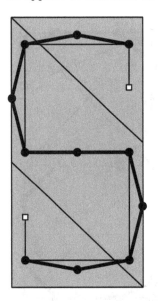

We merely have to scale up the convex hull by a factor of l. To be more precise, each band corresponding to a face orientation needs to be scaled up about its centroid by a factor of l.

Thus enclosures can indeed be found for schemes which are not positive ones.

The straightforward way of determining a good approximation to the l_∞ norm of the scheme is to raise its matrix to a high power and scan down the rows of the resulting matrix. This is not, in fact, particularly onerous, though hardly elegant. Repeated squaring gives high powers very quickly.

The example here is the four-point scheme

$$[-1, 0, 9, 16, 9, 0, -1]/16,$$

for which the l_∞ norm is just over 1.25

The basis is

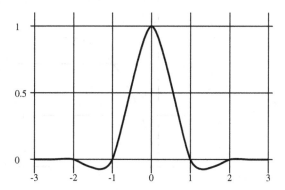

and graphing the copies of the basis functions within one span gives

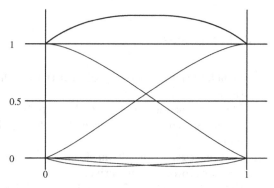

The l_∞ norm graphed over one span is symmetric, and it looks as though the maximum is at the central point of a span. In fact it is not quite. $abs(f(x))$ is itself a function, with V-shaped discontinuities of slope at the roots of $f(x)$.

Expanding the vertical scale of the above plot makes the first span of the basis of the four point scheme visible. It has roots at $\dots, 1/8, 1/4, 1/2$

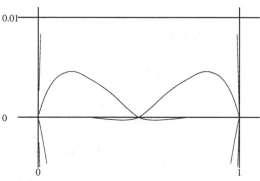

and so the l_∞ norm has V-shaped kinks at

$$\dots, 1/16, 1/8, 1/4, 1/2, 3/4, 7/8, 15/16, \dots$$

and the one at $1/2$ means that the maximum value is displaced slightly, and is slightly greater than the central value of 1.25.

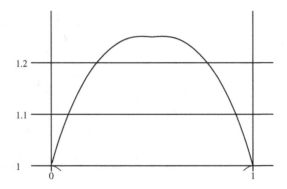

13.3 Exercises

(i)   Which of the five schemes of 11.9(i) above have non-negative basis functions ?

(ii)   Identify a sequence of control points lying within the band $-1 < y < +1$, for which the limit curve of the four-point scheme goes outside that band.

(iii) Is it necessary for all mask entries to be non-negative for the basis function to be non-negative ?

(iv)   Extend the routine for plotting approximate basis functions (exercise (iv) of the previous chapter), to report an approximation to the factor by which the distances between opposite faces of the hull of the control points must be scaled up to give an enclosure for the limit curve.

13.4 Summary

(i)   Enclosures can be found for any convergent[18] scheme. This is easiest if the mask has only positive entries, when the enclosure for any piece can be taken as the enclosure of the control points influencing that piece.

(ii)   However, even when the positivity condition is not met, the l_∞ norm of the basis gives a factor by which the convex hull must be expanded to give a usable enclosure. This norm is thus a measure of the enclosability of the scheme.

---

[18]The question of whether a scheme is convergent or not will be addressed in chapter 16 below.

14. Continuity 1 - at Support Ends

Promenade

In a previous chapter we saw that we should expect to see discontinuities of some derivative in the limit curve at places corresponding to the ends of the support region.

This raises the obvious question '*a discontinuity of what derivative ?* '.

The more general question '*what is the Hölder continuity of the limit curve ?* ' has attracted an enormous amount of attention. Much more indeed than is justified by the importance of the answer to applications. However, some really sharp tools have been developed, which are applicable to addressing the more important issues, and so that work has been well-justified. It also makes a good story, which will be told in the next few chapters. Our two subjects are eigenanalysis and difference schemes. These are developed in counterpoint by looking at the eigenanalysis of difference schemes and the joint spectral radius of a scheme, and brought to a coda by the question of what it is that actually converges to the limit curve.

14.1 Derivative Continuity of the Basis Function at its Ends

We look first at the original question '*What is the level of derivative continuity at the ends of the basis function*'.

This can be addressed informally by looking at the rate at which the final entry of the mask reduces, compared with the rate at which the gap between the last control and the end of the support reduces.

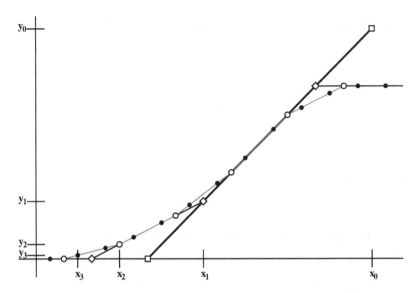

If the final non-zero entry in the mask has the value f and the arity is a, y_j is the final non-zero in the j-times refined polygon (so that $f = y_0$) and x_j is the distance from that entry to the end of the support then $y_j = y_0^j$ and $x_j = s(a^{-j})$, where s is half the support width, the initial value of x (so that $s = x_0$).

We are looking for behaviour of the form $|y| = \alpha x^k$ and this is

$$|y_j| = |y_0|^j$$
$$\alpha x_j^k = \alpha s^k a^{-jk}$$
$$(|y_0|a^k)^j = \alpha s^k$$

The right hand side is constant, independent of j, and so the left hand side must be also. For this to be the case we must have $|y_0|a^k = 1$, or $|y_0| = a^{-k}$. Thus $k = -\log_a(|y_0|) = -\log(|y_0|)/\log(a) = -\log_a(|f|)$

Then we expect derivatives lower than the kth to be continuous at the end of the support region, and the next derivative to have a discontinuity.

This is at first sight a plausible argument, but as it stands it has a number of glaring holes.

•   The values which are converging at a certain rate towards zero are not samples from the limit curve, but control values.

•   There is no guarantee that samples taken in between the dyadic places sampled here are equally well behaved.

•   Although the basis function itself may have a discontinuity of the computed level at its end, the other translates of the basis function may have worse discontinuities in their interiors at the same point, and so this does not tell us anything about the continuity of limit curves in general.

- Worse, this is telling us only about the points corresponding to basis function end-points, not about other points on the limit curve.

Consider these challenges one at a time.

14.1.1 Why does the limit curve converge just because the control points do?

This problem is not hard to fix. It is clear that there is a finite piece of the basis function whose size depends only on the final value in the mask. The influence of the last value reaches to the end of the support, and the influence of the rest does not reach so far. The detail of the shape within that piece may depend on all the values, but the size relative to that of the same piece at other levels of refinement depends only on the end value. It shrinks by a factor of f, just like the end value itself.

The next one along cannot have any influence over the neighbourhood of the end of the support, and that is a finite neighbourhood, not an infinitesimal one.

Within that neighbourhood, the value of the limit function depends only on the extreme mask value, in exactly the same way that the value of the extreme control value does. Those values, which are values of the basis function, scale down from one refinement to the next by exactly the same factors as the control values that we looked at.

14.1.2 How do you know something nasty doesn't happen at places in between the extreme control points examined?

Here we can make a nice illustration of the problem. Consider the function

$$f(x) = \begin{cases} 0 & x \le 0 \\ x\,\cos(2\pi \log_2 x) + \sin(2\pi \log_2 x) & x > 0 \end{cases}$$

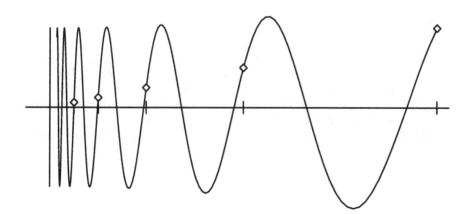

If we sample this at values $x = 1, 1/2, 1/4, 1/8\ldots$, because $\log_2(x)$ is an integer, and so $\sin(2\pi \log_2(x)) = 0$, only the first term contributes. The values are $1, 1/2, 1/4, 1/8\ldots$, which appear to be converging nicely as x.

But at intermediate places there is no convergence to zero.

However, this illustration was not of a subdivision construction. If a subdivision scheme is convergent, we have the result from the previous chapter, that the limit function cannot take a larger value within a span than a certain factor times the largest of a few local control values. By the same argument as for the previous problem, these local control values are all scaling down at the same rate, and so the upper bound on the limit values must also.

Thus the first two of these gaping holes can be plugged: they have reasonable answers. The other two are indeed problems, which will be addressed in later chapters.

Indeed, we can make a subdivision scheme (the mask is $[1,8,14,8,1]/16$) which does indeed have a poorer level of continuity in the limit curve, even at points corresponding to the original control points, than appears at the ends of the basis function.

We can also make a subdivision scheme (the mask is $[2,7,10,7,2]/14$) which has a poorer level of continuity in the curve as a whole than it does at the limit curve points which correspond to control points.

Thus the analysis of this chapter has indeed determined the Hölder continuity, but only of the basis function and only at its end points. There is much more to do before we have a full answer to the continuity question.

14.2 Exercises

(i)   Applying the methods of this chapter to the schemes in 11.9(i) above is totally trivial, but do it anyway.

14.3 Summary

(i)   The Hölder continuity at the ends of the basis function is determined from the value, f, at the ends of the mask and the arity, a. Let $k = -\log_a(|f|) = -\log(|f|)/\log(a)$. Then the highest derivative continuous at the ends of the basis function is $d = \lceil k \rceil - 1$ and the fractional part of the Hölder continuity is $k - d$.

(ii)   This is only an upper bound on the Hölder continuity of the limit curve.

15. Continuity 2 - Eigenanalysis

Promenade

The earliest work on continuity focussed on the questions *" here is a scheme: is it continuous ? is it C^1 ? is it C^2 ?..."* which can be subsumed into the single question *"How many derivatives are continuous ?"*. We now phrase the question a different way: *what are upper and lower bounds on the Hölder continuity ?* This is because numbers like $-\log(|f|)/\log(a)$ are very rarely integers.

The number found in the previous chapter is a strict upper bound on the Hölder continuity because we have an example of a discontinuity at one particular place of one particular limit curve.

In fact, because other places could have discontinuities of lower derivatives and often do, this is usually a rather sloppy upper bound, and the next few chapters deal with ways of finding tighter bounds, both upper and lower. When the bounds converge, we can say that we know the continuity of the limit curve.

15.1 Continuity at Mark Points by Eigenanalysis

Because each vertex of the refined control polygon is a weighted mean of vertices of the original, the construction of a refined control polygon can be expressed in the form

$$P' = S P$$

where P is a column vector whose entries are the vertices of the initial polygon, S is a matrix all of whose rows sum to 1, and P' is a column vector holding the new polygon.

S is called the **Subdivision Matrix**. It has an interesting structure. Every column is a copy of the mask, but successive columns have their copies shifted down by the arity compared with their left neighbours. For example, the subdivision matrix for the Cubic B-spline scheme is

M. Sabin, *Analysis and Design of Univariate Subdivision Schemes*, Geometry and Computing 6,   81
DOI 10.1007/978-3-642-13648-1_15, © Springer-Verlag Berlin Heidelberg 2010

$$
\begin{bmatrix}
\ddots & & & & & & & \\
& 1 & 6 & 1 & & & & \\
& & 4 & 4 & & & & \\
& & 1 & 6 & 1 & & & \\
& & & 4 & 4 & & & \\
& & & 1 & 6 & 1 & & \\
& & & & 4 & 4 & & \\
& & & & 1 & 6 & 1 & \\
& & & & & 4 & 4 & \\
& & & & & 1 & 6 & 1 \\
& & & & & & & \ddots
\end{bmatrix} \times \frac{1}{8}
$$

Note that because we are ignoring what happens at the ends of the polygon, this has to be treated as an infinite matrix, and we have to be very careful to justify steps which are only known to apply to finite matrices.

The interesting property of an infinite matrix is that, lacking a top left hand corner, it doesn't have a principal diagonal. Any diagonal can be taken as principal. Because we have chosen a binary scheme for the example, all diagonals look the same anyway, so it makes no difference for our analysis which one we choose.

What the choice of a diagonal does is to imply a labelling, giving a correspondence between a sequence of points of the old polygon and a sequence of the refined one. In particular it implies a **mark point** which is an abscissa value which maps into itself under the map from old abscissa values to new ones. In the case of a primal binary scheme, the mark point is at a point of both new and old polygons. In the case of a dual scheme the mark point is at a mid-edge in both old and new.

Because the slope of the non-zero entries in the matrix is greater than 1, it also defines a square region of the matrix for which no point outside the row-range of that square influences any new point in the column-range.

$$
8\begin{bmatrix}
\vdots \\
w \\
x \\
y \\
z \\
a \\
b \\
c \\
d \\
e \\
\vdots
\end{bmatrix}
=
\begin{bmatrix}
\ddots & & & & & & \\
1 & 6 & 1 & & & & \\
& 4 & 4 & & & & \\
& 1 & 6 & 1 & & & \\
& & 4 & 4 & & & \\
& & 1 & 6 & 1 & & \\
& & & 4 & 4 & & \\
& & & 1 & 6 & 1 & \\
& & & & 4 & 4 & \\
& & & & 1 & 6 & 1 \\
& & & & & & \ddots
\end{bmatrix}
\begin{bmatrix}
\vdots \\
W \\
X \\
Y \\
Z \\
A \\
B \\
C \\
D \\
E \\
\vdots
\end{bmatrix}
$$

Thus W, X, D and E do not affect any of y,z,a,b or c.

Y to C are exactly the control points which affect the neighbourhood of A, as determined above, in the support chapter.

We may therefore take just this square region, and forget the rest when we are looking at the continuity of the limit curve at A. When we multiply the Y to C part of the initial polygon by the matrix repeatedly, we construct refined polygons which in the limit converge to a tiny neighbourhood of the limit curve, and it is that tiny neighbourhood which determines the continuity at A. Eigenanalysis tells us what happens when we multiply repeatedly.

5×5 is a rather large matrix to address directly by hand, but we can apply both symmetry and the block structures of the symmetry-partition matrices to make the calculations easy.

15.1.1 Odd-even partitioning

The sequence Y, Z, A, B, C can be expressed as the sum of two components

$$8 \begin{bmatrix} Y \\ Z \\ A \\ B \\ C \end{bmatrix} = \begin{bmatrix} (C+Y)/2 \\ (B+Z)/2 \\ A \\ (B+Z)/2 \\ (C+Y)/2 \end{bmatrix} + \begin{bmatrix} -(C-Y)/2 \\ -(B-Z)/2 \\ 0 \\ (B-Z)/2 \\ (C-Y)/2 \end{bmatrix}$$

where the first component is symmetric and the second is antisymmetric.

Because of the symmetry of the 5×5 matrix, when we multiply the first component by the matrix we get a symmetric result, and when we multiply the second component by the matrix we get an antisymmetric result. We can thus look at each of these cases.

$$8 \begin{bmatrix} a \\ (b+z)/2 \\ (c+y)/2 \end{bmatrix} = \begin{bmatrix} 6 & 2 & \\ 4 & 4 & \\ 1 & 6 & 1 \end{bmatrix} \begin{bmatrix} A \\ (B+Z)/2 \\ (C+Y)/2 \end{bmatrix}$$

$$8 \begin{bmatrix} (b-z)/2 \\ (c-y)/2 \end{bmatrix} = \begin{bmatrix} 4 & \\ 6 & 1 \end{bmatrix} \begin{bmatrix} (B-Z)/2 \\ (C-Y)/2 \end{bmatrix}$$

Note that these two linear systems together have enough eigencomponents to equal the number of eigencomponents of the original 5×5. We have not ignored any mixed eigencomponents. Indeed, the simple counting is a proof that for a palindromic binary primal subdivision scheme every eigencomponent must have either a symmetric or an antisymmetric column eigenvector. A similar proof applies to dual schemes, where the number of symmetric components is the same as the number of antisymmetric ones.

15.1.2 Using block structure

The next simplification of the calculations is by exploiting the block structure of these matrices. We use the property that if a matrix is block lower triangular, its eigenvalues are those of the diagonal blocks.

The eigenvalues of the upper, symmetric, case are given by those of $\begin{bmatrix} 6 & 2 \\ 4 & 4 \end{bmatrix}$, which are 8 and 2, and 1 from the trailing block. The unnormalised eigencolumns are

$$\begin{bmatrix} 1 \\ 1 \\ 1 \end{bmatrix} \quad \begin{bmatrix} -1 \\ 2 \\ 11 \end{bmatrix} \quad \begin{bmatrix} 0 \\ 0 \\ 1 \end{bmatrix}$$

and the unnormalised eigenrows are

$$\begin{bmatrix} 2 & 1 & 0 \end{bmatrix}$$
$$\begin{bmatrix} -1 & 1 & 0 \end{bmatrix}$$
$$\begin{bmatrix} 0 & 0 & 1 \end{bmatrix}.$$

Those of the lower, antisymmetric, case are 4 and 1 from the lower triangular block structure.

The unnormalised eigencolumns are

$$\begin{bmatrix} 1 \\ 2 \end{bmatrix} \quad \begin{bmatrix} 0 \\ 1 \end{bmatrix}$$

and the unnormalised eigenrows are

$$\begin{bmatrix} 1 & 0 \end{bmatrix}$$
$$\begin{bmatrix} 2 & -1 \end{bmatrix}$$

Observe that the matrices containing the eigenvectors have the same lower triangular block structure as the originals.

15.1.3 Interpretation

We now need to divide by the denominator 8, previously tucked away on the left hand sides of the equations, to find the complete set of eigenvalues of the original matrix

$$1, 1/2, 1/4, 1/8, 1/8,$$

and use the original construction of the symmetric and antisymmetric partitions to determine the eigenvectors.

The unnormalised eigencolumns are

$$\begin{array}{ccccc} 1 & 1/2 & 1/4 & 1/8 & 1/8 \end{array}$$

$$\begin{bmatrix} 1 \\ 1 \\ 1 \\ 1 \\ 1 \end{bmatrix} \quad \begin{bmatrix} -2 \\ -1 \\ 0 \\ 1 \\ 2 \end{bmatrix} \quad \begin{bmatrix} 11 \\ 2 \\ -1 \\ 2 \\ 11 \end{bmatrix} \quad \begin{bmatrix} -1 \\ 0 \\ 0 \\ 0 \\ 1 \end{bmatrix} \quad \begin{bmatrix} 1 \\ 0 \\ 0 \\ 0 \\ 1 \end{bmatrix}$$

and the unnormalised eigenrows are

$$
\begin{array}{cc}
1 & \begin{bmatrix} 0 & 1 & 4 & 1 & 0 \end{bmatrix} \\
1/2 & \begin{bmatrix} 0 & -1 & 0 & 1 & 0 \end{bmatrix} \\
1/4 & \begin{bmatrix} 0 & 1 & -2 & 1 & 0 \end{bmatrix} \\
1/8 & \begin{bmatrix} -1 & 2 & 0 & -2 & 1 \end{bmatrix} \\
1/8 & \begin{bmatrix} 1 & -4 & 6 & -4 & 1 \end{bmatrix}
\end{array}
$$

By taking larger square matrices with initial and final columns of zeroes, we find that the eigencolumns are just parts of much longer non-zero columns.

$$
\begin{array}{ccccc}
1 & 1/2 & 1/4 & 1/8 & 1/8
\end{array}
$$

$$
\begin{bmatrix} \vdots \\ 1 \\ 1 \\ 1 \\ 1 \\ 1 \\ 1 \\ 1 \\ \vdots \end{bmatrix}
\begin{bmatrix} \vdots \\ -3 \\ -2 \\ -1 \\ 0 \\ 1 \\ 2 \\ 3 \\ \vdots \end{bmatrix}
\begin{bmatrix} \vdots \\ 26 \\ 11 \\ 2 \\ -1 \\ 2 \\ 11 \\ 26 \\ \vdots \end{bmatrix}
\begin{bmatrix} \vdots \\ -8 \\ -1 \\ 0 \\ 0 \\ 0 \\ 1 \\ 8 \\ \vdots \end{bmatrix}
\begin{bmatrix} \vdots \\ 8 \\ 1 \\ 0 \\ 0 \\ 0 \\ 1 \\ 8 \\ \vdots \end{bmatrix}
$$

but the eigenrows are not extended in this way. They stay short. This ties up with the fact that the number of original control points influencing a neighbourhood of the limit curve is limited by the support arguments above. A larger matrix does, of course, have additional eigencomponents, but these all have zero eigenvalues, and so we can ignore them, at least for the next few chapters.

Clearly the eigenvalue 1 is dominant. Its column eigenvector in the original matrix is a column of all 1s[19], which means that all the points in this piece of the control polygon will be at the same place in the limit.

That place is given by multiplying the original polygon by the normalised eigenrow, which gives $(Z + 4A + B)/6$.

This turns out to be an extremely important result. Being able to construct limit points without actually doing an infinite number of refinements is a key to practical use of subdivision curves and surfaces.

In order to look more closely at the neighbourhood of this point, let us choose the coordinate system[20] in which the initial polygon is expressed, so that this point is at the origin. There is then no contribution to this

---

[19]In fact as long as each of the stencils sums to 1, the matrix will always have a unit eigenvalue with an eigencolumn of 1s

[20]Can we do this ? Yes, because each stencil defines an affine combination, and affine combinations are invariant under translations. In fact they are invariant under solid body rotations, scalings and affine transforms too.

eigencomponent, and we can see, by looking at the next eigencomponent, what is happening near the origin.

The next column eigenvector is varying linearly, and so the polygon is converging towards a straight line, with points evenly spaced along it. The eigenvalue is $1/2$, which means that the density doubles at each step. The first derivative at the limit point is the limit of the first divided difference, which is $(B - Z)/2$.

To explore further, we choose our coordinate system so that this straight line is the x-axis. The subdominant eigencomponent then makes no contribution to y, which is dominated by the third eigencomponent. The column eigenvector looks complicated, but in fact it is just a quadratic variation, with an offset added

$$
\begin{bmatrix} \vdots \\ 26 \\ 11 \\ 2 \\ -1 \\ 2 \\ 11 \\ 26 \\ \vdots \end{bmatrix}
=
\begin{bmatrix} \vdots \\ 27 \\ 12 \\ 3 \\ 0 \\ 3 \\ 12 \\ 27 \\ \vdots \end{bmatrix}
-
\begin{bmatrix} \vdots \\ 1 \\ 1 \\ 1 \\ 1 \\ 1 \\ 1 \\ 1 \\ \vdots \end{bmatrix}
= 3
\begin{bmatrix} \vdots \\ 9 \\ 4 \\ 1 \\ 0 \\ 1 \\ 4 \\ 9 \\ \vdots \end{bmatrix}
-
\begin{bmatrix} \vdots \\ 1 \\ 1 \\ 1 \\ 1 \\ 1 \\ 1 \\ 1 \\ \vdots \end{bmatrix}
$$

This component is giving the curvature of the limit curve. The second derivative at the limit point is given by the limit of the second divided difference, which is $B + Z - 2A$.

We can keep exploring, by again choosing our coordinate system so that this component lies in the xy-plane, and then look at the z-values.

Now we have two components with the same eigenvalue, $1/8$.

The antisymmetric one follows exactly the same pattern as before, giving us a well-defined third derivative, given by the mean, $C - 2B + 2Z - Y$, of the third divided differences to left and right.

However, the symmetric component gives something more awkward. The limit has different third divided differences on left and right, which means that there is a discontinuity of third derivative. The size of this discontinuity is given by the fourth divided difference of the original polygon.

Again we have constructed a discontinuity which can be used as an upper bound, in this case 2+1, on the Hölder continuity.

Because we could have carried out the analysis after applying a few levels of refinement, the discontinuity found in this way can, in principle, be found at limit points corresponding to control points at any level of refinement. These points are called **dyadic points**. They are dense, but not as dense as the rationals, let alone the reals. Between any two of them, however close, you can find a point which is not a dyadic.

This procedure can be applied to any primal binary scheme, although it may be necessary to imagine higher dimensions than 3 in order to keep applying the principle of suppressing successive dominant eigencomponents.

In the particular case shown as an example here, at the mark points constructed in this way other than at abscissae of original vertices, the fourth difference turns out to be zero, and so the discontinuities do not occur except at limit points corresponding to original control points. That ties up with our knowledge of B-splines, but it is a very special property. In general, subdivision schemes give limit curves with discontinuities of some derivative at all dyadic points.

Now consider the scheme whose mask is $[1, 8, 14, 8, 1]/16$. Clearly the discontinuity at the ends is $-\log_2(1/16) = 4$. The basis function is C^{3+1} there. However, eigenanalysis gives a different result.

The matrix is

$$\begin{bmatrix} 1 & 14 & 1 & & \\ & 8 & 8 & & \\ & 1 & 14 & 1 & \\ & & 8 & 8 & \\ & & 1 & 14 & 1 \end{bmatrix} /16$$

The symmetric and antisymmetric components are

$$\begin{bmatrix} 14 & 2 & \\ 8 & 8 & \\ 1 & 14 & 1 \end{bmatrix} /16 \quad \text{and} \quad \begin{bmatrix} 8 & \\ 14 & 1 \end{bmatrix} /16$$

The symmetric eigenvalues are $1, 6/16, 1/16$ and the antisymmetric ones $1/2, 1/16$. The complete set is therefore $[1, 1/2, 3/8, 1/16, 1/16]$ which compares with the $[1, 1/2, 1/4, 1/8, 1/8]$ of the cubic B-spline scheme. The important difference is the $3/8$ which replaces the $1/4$.

The unnormalised eigenvectors of the first three eigenvalues are

$$\begin{bmatrix} \vdots \\ 1 \\ 1 \\ 1 \\ 1 \\ 1 \\ 1 \\ 1 \\ \vdots \end{bmatrix} \begin{bmatrix} \vdots \\ -3 \\ -2 \\ -1 \\ 0 \\ 1 \\ 2 \\ 3 \\ \vdots \end{bmatrix} \begin{bmatrix} \vdots \\ 20 \\ 11 \\ 4 \\ -1 \\ 4 \\ 11 \\ 20 \\ \vdots \end{bmatrix}$$

The first two of these are well-behaved, but the third is definitely not quadratic in shape[21]. We can therefore say that the largest non-polynomial

---

[21]If more of it is evaluated it turns out that it is not polynomial at all.

eigencomponent has eigenvalue $3/8$, and that the Hölder continuity is no higher than $-\log_2(3/8) \approx 1.418$.

$C^{1+0.418}$ is a significantly lower continuity level than the C^{3+1} which our endpoint analysis gave. The endpoint has the better continuity level because the more typical behaviour has as its amplitude the product of the local configuration with the relevant row eigenvector, which in this case is $[1, -2, 1]$, which is zero two control points away from the 1 in cardinal data.

15.2 A Motivation Question

Why do all the arithmetic of calculating eigencomponents, when there is software available to do it for us ?

The reason for this tedious working through is that the separations we have made by symmetry and by block structure do enable us to pick out and observe patterns in the eigenvectors which could be confused when there are two or more eigencomponents with the same eigenvalue. These patterns are going to be significant in a couple of chapters' time, and it is important to observe them empirically first.

15.3 Dual Schemes

In the case of dual schemes, the piece of matrix which has to be dealt with has an even number of rows and columns. The partitioning into even and odd, symmetric and antisymmetric partitions is therefore slightly neater. It also means that the place where the continuity is assessed (like the ends of the support) is halfway between control points rather than being at them. Otherwise the process is exactly analogous. The partitioning and the use of block structure works in exactly the same way.

15.4 Higher Arities

The slope of the subdivision matrix is equal to the arity, and so when we have a scheme with an arity higher than two, not all diagonals are equal. We find $a - 1$ different possible matrices to analyse. For $a = 3$ we get analyses at limit points corresponding to the control points and to the middles of the spans. In each case the choice of a diagonal identifies a labelling, which can then be viewed as defining a centre of symmetry.

$$
3 \begin{bmatrix} a \\ b \\ c \end{bmatrix} = \begin{bmatrix} & 3 & \\ 2 & 1 & \\ 1 & 2 & \\ & 3 & \\ & 2 & 1 \\ & 1 & 2 \\ & & 3 \end{bmatrix} \begin{bmatrix} A \\ B \\ C \end{bmatrix}
$$

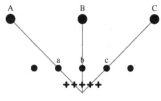

$$
3 \begin{bmatrix} a \\ b \\ c \\ d \end{bmatrix} = \begin{bmatrix} & 3 & \\ & 2 & 1 \\ & 1 & 2 \\ & 3 & \\ & 2 & 1 \\ & 1 & 2 \\ & & 3 \end{bmatrix} \begin{bmatrix} A \\ B \\ C \\ D \end{bmatrix}
$$

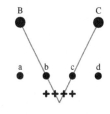

For $a = 4$, we get something much more interesting, points one third of the way along a span, which are not points which ever get explicitly constructed by the refinement.

$$
3 \begin{bmatrix} a \\ \mathbf{b} \\ \mathbf{c} \\ d \end{bmatrix} = \begin{bmatrix} 2 & 2 & \\ 1 & 3 & \\ & 4 & \\ 3 & 1 & \\ 2 & 2 & \\ 1 & 3 & \\ & 4 & \end{bmatrix} \begin{bmatrix} A \\ \mathbf{B} \\ \mathbf{C} \\ D \end{bmatrix}
$$

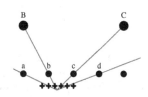

It then turns out that the discontinuities found at such points may be totally different from those at the dyadic points. We can always make schemes of higher arity by considering two or more refinements as a single step. We call this **squaring** or **taking a higher power of** the scheme.

The binary scheme $[2, 7, 10, 7, 2]/14$ illustrates this.

The matrix is

$$
\begin{bmatrix} 2 & 10 & 2 & & \\ & 7 & 7 & & \\ & 2 & 10 & 2 & \\ & & 7 & 7 & \\ & & 2 & 10 & 2 \end{bmatrix} /14
$$

the symmetric and antisymmetric parts are

$$
\begin{bmatrix} 10 & 4 & \\ 7 & 7 & \\ 2 & 10 & 2 \end{bmatrix} /14 \quad \text{and} \quad \begin{bmatrix} 7 & \\ 10 & 2 \end{bmatrix} /14
$$

and the eigenvalues $1, 1/2, 3/14, 2/14, 2/14$. The components with eigenvalues 1 and 1/2 are perfectly well-behaved, and so the focus is on the 3/14 value, which is less than 1/4.

If, however, we take the square of the scheme, we get a quaternary scheme

$$[4, 14, 34, 63, 94, 119, 128, 119, 94, 63, 34, 14, 4]/196$$

whose matrix is

$$\begin{bmatrix}
4 & 94 & 94 & 4 & & & & & \\
 & 63 & 119 & 14 & & & & & \\
 & 34 & 128 & 34 & & & & & \\
 & 14 & 119 & 14 & & & & & \\
 & 4 & 94 & 94 & 4 & & & & \\
 & & 63 & 119 & 14 & & & & \\
 & & 34 & 128 & 34 & & & & \\
 & & 14 & 119 & 14 & & & & \\
 & & 4 & 94 & 94 & 4 & & & \\
 & & & 63 & 119 & 14 & & & \\
 & & & 34 & 128 & 34 & & & \\
 & & & 14 & 119 & 14 & & & \\
 & & & 4 & 94 & 94 & 4 & & \\
 & & & & 63 & 119 & 14 & & \\
 & & & & 34 & 128 & 34 & & \\
 & & & & 14 & 119 & 14 & & \\
 & & & & 4 & 94 & 94 & 4 &
\end{bmatrix} /196$$

If we choose a diagonal through the value 128, we find a 5×5 matrix which is just the square of that of the original scheme. Its eigenvalues are therefore the squares of those determined above, and the Hölder continuiity would appear to be better than 2.

A diagonal chosen one higher however gives a 4×4 matrix

$$\begin{bmatrix}
14 & 119 & 14 & \\
4 & 94 & 94 & 4 \\
 & 63 & 119 & 14 \\
 & 34 & 128 & 34
\end{bmatrix} /196$$

which has eigenvalues $[1, 1/4, 0.0678133775, 0.0051020408]$, the third of which is greater than 1/16. The Hölder continuity at this point is only $1+0.9411397$.

In general every power of the scheme considered introduces new mark points, and there is no guarantee in general that the 117th power will not show us places where the Hölder continuity is lower than that found for lower powers. Thus this procedure can only ever give us upper bounds on the Hölder continuity.

15.5 Piecewise Polynomial Schemes

In some rather special cases, it is possible for the matrices on some diagonals to have only polynomial eigencomponents, saying that the Hölder continuity is infinite. This happens for the B-splines.

$$
\begin{bmatrix}
\ddots & 1 & & & & \\
 & 2 & & & & \\
 & 1 & 1 & & & \\
 & & 2 & & & \\
 & & 1 & 1 & & \\
 & & & 2 & \ddots & \\
 & & & 1 & \ddots &
\end{bmatrix}^2
/2^2 =
\begin{bmatrix}
3 & 1 & \\
2 & 2 & \\
1 & 3 & \\
 & 4 & \\
 & 3 & 1 \\
 & 2 & 2 \\
 & 1 & 3
\end{bmatrix} /4
$$

and the small matrix $\begin{bmatrix} 3 & 1 \\ 2 & 2 \end{bmatrix}$ has eigenvalues 1 and 1/4 only. The eigenvectors are polynomial. This is hardly surprising, because we know that the limit curve is piecewise polynomial, having complete continuity in the interior of its spans.

An exactly similar effect happens for the ternary neither schemes. The square of $[1, 3, 5, 5, 3, 1]/6$ is $[1, 3, 5, 8, 12, 16, 20, 24, 28, 30, 30, 30, 28, 24, \ldots]/36$ and within its matrix (arity 9, so there is lots of step down from one column to the next) we find the submatrix

$$
\begin{bmatrix}
20 & 16 \\
16 & 20
\end{bmatrix} /36
$$

which has eigenvalues 1 and 1/9, with polynomial eigenvectors. Again this ties up with what we know from the support arguments in section 12.3.

15.6 What Mark Points can be Made ?

By taking a high enough power of the scheme, any rational point can be determined as a mark point. The power needed is just the Euler function of the quotient when the denominator has all powers of 2 (in general of the arity) divided out.

However, at the time of writing, '*What is the simplest scheme with a Hölder-dominant denominator greater than 3*' was still an open question.

The first few powers and the denominators of the mark points that they can find are

Power	Arity	No. of Matrices	Denominator factors
1	2	1	1
*2	4	3	3
3	8	7	7
*4	16	15	$3, 5, 15$
5	32	31	31
6	64	63	$3, 7, 9, 27, 63$
7	128	127	127
*8	256	255	$3, 5, 15, 17, 51, 85, 255$
9	512	511	$7, 73, 511$
10	1024	1023	$3, 11, 31, 33, 93, 341, 1023$
11	2048	2047	$23, 89, 2047$
12	4096	4095	$3, 5, 7, 9, 13, 15, 21, 39, 63, \ldots, 4095$
13	8192	8191	8191
14	16384	16383	$3, 43, 127, 129, 381, \ldots, 5461, 16383$
15	32768	32767	$7, 31, 217, 151, 1057, \ldots, 32767$
*16	65536	65535	$3, 5, 15, 17, 51, 85, 255, 257, \ldots, 65535$

The powers noted by * are those cheaply created by successive squaring. Clearly some powers are richer than others for covering many denominators.

15.7 Exercises

One of the five schemes in 11.9(i) has already been analysed in detail above. The exercises here are essentially about eigenanalyses for the others, taking in the complications mentioned in this chapter.

(i)   Find the Hölder continuity of the quadratic B-spline scheme [1,3,3,1]/4 at the centre of the spans between control points.

(ii)   Find the Hölder continuity at both of the mark points of the ternary quadratic scheme [1,3,6,7,6,3,1]/9 .

(iii) Find the Hölder continuity of the ternary neither scheme scheme [1,3,5,5,3,1]/6 at the 1/4 point.

(iv)   Find the Hölder continuity of the four-point scheme [-1,0,9,16,9,0,-1] at the limit points corresponding to the control points.

(va)   Write a routine to do this analysis for any given mask. You are advised to find a library routine (rather than writing your own) to do the actual eigenanalysis, so all you have to do is form the appropriate matrix/matrices to provide the input arguments for that routine. Don't worry about the short cuts if you have a library routine to handle big matrices. Remember that for arities greater than 2 there can be more than one matrix to handle.

(vb)   Smarten that routine up by applying the short cuts before calling the library routine on smaller matrices, and compare the results. A good library routine will produce symmetric and antisymmetric eigenvectors when there is a repeated eigenvalue, but a less good one may not. In any case reducing the size of the matrices cannot make the precision of the results any worse.

15.8 Summary

(i)   This chapter has been full of tedious arithmetic. However, it illustrates a procedure which can be applied systematically to any scheme to determine the continuity at mark points, which are at control points in the case of primal binary schemes and at mid-edges in the case of dual ones. If we did not have better procedures, to be described in the next three chapters, it could be programmed as an algorithm, taking as input only the arity and the mask.

(ii) This level of continuity applies at mark points after any number of refinements (because we could just have done those refinements before starting the analysis). These points are called the **dyadic points**, and they are dense.

(iii) However, there are other points in between, which may have different continuity properties, and so we can only achieve upper bounds in this way. Tighter upper bounds can be found by taking powers of the scheme, which make some of these in-between points explicit.

(iv) This procedure can be applied to any scheme, although it may be necessary to imagine higher dimensions than 3 in order to keep applying the principle of suppressing successive dominant eigencomponents.

(v)   The row eigenvalues of unit eigenvalue (the **unit eigenrows**) can give explicit stencils for any rational parameter value. These stencils allow evaluation of points on the limit curve itself with a relatively small amount of calculation.

16. Continuity 3 - Difference Schemes

Promenade

We have seen that it is possible to place upper bounds on the continuity of
a scheme by carrying out eigenanalysis around a mark point. In principle
these upper bounds can be tightened by doing this analysis for powers of the
scheme, which give additional markpoints.

However, if we are ever to say with confidence *'This is the Hölder continuity of curves generated by this scheme'*, we need also lower bounds, which
we can approach rigorously by using difference schemes.

16.1 Lower Bounds by Difference Schemes

16.1.1 Continuity by difference schemes

This approach uses the definition of continuity, that a function $f(x)$ is continuous at an abscissa x if

$$Lt_{\delta x \to 0} f(x + \delta x) - f(x) = 0$$

The sequence of δx used for taking this limit is conveniently the sequence
of polygon edges at successive refinements of the original polygon.

The question is whether we can bound the values of $f(x + \delta x) - f(x)$ in
terms of the original control points, and the answer is 'yes', using the neat
idea of a **difference scheme**, which relates the first differences of the new
polygon to the first differences of the old.

Suppose that a binary scheme, S, has a z-transform of $S(z)$ and the old
polygon is $P_0(z^2)$. Then the new polygon is given by

$$P_1(z) = S(z)P_0(z^2)$$

Now the generating function of the first differences δP_0 of P_0 is just $(1 -
z^2)P_0(z)$ and that of the first differences δP_1 of the new one $(1 - z)P_1(z)$.

M. Sabin, *Analysis and Design of Univariate Subdivision Schemes*, Geometry and Computing 6,
DOI 10.1007/978-3-642-13648-1_16, © Springer-Verlag Berlin Heidelberg 2010

$$\delta P_1 = (1 - z)P_1(z)$$
$$= (1 - z)S(z)P_0(z^2)$$
$$= \frac{1 - z^2}{1 + z}S(z)P_0(z^2)$$
$$= \frac{S(z)}{1 + z}(1 - z^2)P_0(z^2)$$
$$= \frac{S(z)}{1 + z}\delta P_0$$

Thus if $S(z)$ is divisible by $1 + z$ (and it always is if each of the stencils sums to 1) we can take the quotient as a scheme which relates first differences of P_1 to first differences of P_0. Call this scheme $D(z)$.

Now if for any sequence δP_0 the corresponding sequence δP_1 has all its entries strictly smaller than the largest entry in δP_0, then the largest first difference shrinks at every step and so we prove that the limit curve is continuous.

That will be the case if the sum of the absolute values of the entries in each row[22] of the subdivision matrix (alternate entries in the mask of $D(z)$) is strictly less than one. If this is the case we say that the scheme is **contractive**.

There is an apparent hole in this argument. We are using only dyadic points for looking at the convergence. How do we know that all differences in between also contract ? As in section 14.1.2 above, the answer is in the enclosure property. If the first differences contract, then a long enough sequence of consecutive first differences to define a span must also contract, and enclosure then says that the span itself must contract[23].

Take the example of the cubic B-spline again. The mask is $[1,4,6,4,1]/8$ so the symbol of the mask is $S(z) = (1 + 4z + 6z^2 + 4z^3 + z^4)/8$, and if we divide this by $1 + z$ we get $D(z) = (1 + 3z + 3z^2 + z^3)/8$. The rows of its matrix are all copies of either $[1, 3]/8$ or $[3, 1]/8$ and the sum of the absolute values is in each case $4/8 = 1/2 < 1$.

This shows that the limit curve is continuous, not just at a mark point, or at dyadic points, but at all points of the curve.

In contrast we can take the scheme whose mask is $[-1,5,5,-1]/4$, and whose symbol is $(-1 + 5z + 5z^2 - z^3)/4$. The difference scheme has symbol $(-1 + 6z - z^2)/4$. The rows (stencils) are $(-1, -1)/4$ and $6/4$. The row sums are therefore $2/4$ and $6/4$. The second of these is greater than 1 and so there is no guarantee that the limit curve is continuous. In fact it diverges nicely.

---

[22]This is the l_∞ norm

[23]This is a circular argument, because we used continuity in developing the enclosure property. This will eventually get resolved properly in chapter 19.

16.2 Continuity of Derivatives by Divided Difference Schemes

The question of continuity of the limit function itself was addressed by using difference schemes. Closely related is the **divided difference scheme**, which is based on the definition of the derivative at a point.

$$df/dx = Lt_{\delta x \to 0} \frac{f(x + \delta x) - f(x)}{\delta x}$$

The change from the definition of convergence to this definition is essentially a factor of δx in the denominator. Because at each refinement the δx is just halved (or in general divided by the arity), all we need to do is to take the difference scheme $D(z)$ and multiply it by the arity, giving a scheme which we shall call $T(z)$.

$$T(z) = aD(z) = \left(\frac{a(1-z)}{1-z^a} \right) S(z)$$

We then have a scheme which relates divided differences in P_0 to divided differences in P_1, and which, if it converges, converges to the derivative function.

The test for whether it converges is just the test described above. Take the difference scheme of $T(z)$ and see whether it is contractive.

In the cubic spline case $T(z) = \frac{1+3z+3z^2+z^3}{4}$. Its difference scheme is $\frac{1+2z+z^2}{4}$ which has rows which are all copies of $[2]/4$ or $[1,1]/4$. The sum of the absolute values of the entries in each row is $2/4 < 1$, and therefore the divided difference scheme converges and the first derivative is continuous.

We can keep going, taking higher and higher divided difference schemes, until we find a scheme whose difference scheme is not contractive. The last scheme which was continuous gives a lower bound on the Hölder continuity of the original scheme.

16.3 Dual Schemes

The procedure is identical for dual schemes. Indeed, if the original scheme is a primal scheme, its first divided difference scheme is a dual scheme.

16.4 Higher Arities

For a higher arity, a, the necessary changes in the procedure are:-
• that the scheme has to be divisible by $1 + z + \ldots + z^{a-1}$ in general, in place of the $1 + z$ which is the particular case of this for $a = 2$, to give the difference scheme.
• that the difference scheme is multiplied by the arity to give the divided difference scheme.

It is convenient to make a common notation for all arities by defining the symbol[24] $\sigma := (1 - z^a)/a(1 - z)$. Then the divided difference scheme is constructed by dividing by σ. It is computationally convenient then to replace the condition that the l_∞ norm of the difference scheme be less than 1 by the equivalent condition that the l_∞ norm of the divided difference scheme be less than the arity.

16.5 Tightening the Lower Bound

However, the bounds determined in this way are only lower bounds. It is possible for a scheme to fail at a certain level, even when the Hölder continuity is actually higher. Just as in the eigenanalysis case, we can often tighten the bounds by taking a power of the scheme.

A good example is the 4-point scheme, whose mask is

$$[-1, 0, 9, 16, 9, 0, -1]/16.$$

The difference scheme is $[-1, 1, 8, 8, 1, -1]/16$ whose largest row sum is $10/16 < 1$ and so the limit curve is continuous.

The first divided difference scheme is $[-1, 1, 8, 8, 1, -1]/8$, whose difference scheme is $[-1, 2, 6, 2, -1]/8$. The largest row sum is $8/8$ which is not strictly less than 1, and so we cannot assert from this that the first derivative is continuous.

If we take two steps together, however, we get a quaternary scheme, which we can refer to as the **square** of the original, whose mask is

$$[1, 0, -9, -16, -18, 0, 66, 144, 216, 256, 216, 144, 66, 0, -18, -16, -9, 0, 1]/256.$$

The difference scheme is now given by dividing by $(1 - z^4)/(1 - z) = 1 + z + z^2 + z^3$ to give

$$[1, -1, -9, -7, -1, 17, 57, 71, 71, 57, 17, -1, -7, -9, -1, 1]/256$$

---

[24]In later chapters, where symmetry is important, we shall use a slightly more complicated version of this, $\sigma = (1 - z^a)/((1 - z)z^{(a-1)/2})$ which has coefficients for positive and negative powers of z equal in pairs.

whose rows are copies of $[1, -1, 71, -7]/256$ or $[-1, 17, 57, -9]/256$, so that the absolute row sums are $80/256$ and $84/256$, confirming that the limit function is continuous.

The divided difference scheme is four times the difference scheme:

$$[1, -1, -9, -7, -1, 17, 57, 71, 71, 57, 17, -1, -7, -9, -1, 1]/64.$$

The difference scheme of the divided difference scheme is

$$[1, -2, -8, 2, 7, 16, 32, 16, 7, 2, -8, -2, 1]/64$$

whose stencils are all copies of $[1, 7, 7, 1]/64$, $[2, 16, -2]/64$, $[-8, 32, -8]/64$ or $[-2, 16, 2]/64$, and so the largest absolute row sum $= 48/64 < 1$. The first divided difference scheme is convergent and so the four-point scheme has a continuous first derivative.

These calculations can be eased significantly by using the facts that

(i)   The difference scheme of the square of a scheme is the square of the difference scheme of the original.

(ii)   The divided difference scheme of the square of a scheme is the square of the divided difference scheme of the original.

Write $S_2(z)$ to denote the scheme one step of which is the same as two steps of $S(z)$, and $T_2(z)$ to denote the scheme one step of which is the same as two steps of $T(z)$.

$$S_2(z) = S(z)S(z^2)$$
$$= \frac{1-z}{1-z^2}T(z)\frac{1-z^2}{1-z^4}T(z^2)$$
$$= \frac{1-z}{1-z^4}T(z)T(z^2)$$
$$= \frac{1-z}{1-z^4}T_2(z)$$

This easing of calculations is illustrated by merely taking the square of $[-1, 2, 6, 2, -1]/8$ to give $[1, -2, -8, 2, 7, 16, 32, 16, 7, 2, -8, -2, 1]/64$ instead of going via $[1, -1, -9, -7, -1, 17, 57, 71, 71, 57, 17, -1, -7, -9, -1, 1]/256$

| -1 | | 2 | | 6 | | 2 | | -1 | | | | |
| -1 | 2 | 6 | 2 | -1 | | | | | | | | |

1		-2		-6		-2		1				
	-2		4		12		4		-2			
		-6		12		36		12		-6		
			-2		4		12		4		-2	
				1		-2		-6		-2		1

| 1 | -2 | -8 | 2 | 7 | 16 | 32 | 16 | 7 | 2 | -8 | -2 | 1 |

Clearly this property can be extended to any number of steps taken at once, and to divided difference schemes.

16.6 A Procedure for Determining Bounds on Hölder Continuity

A systematic procedure, which can be totally automated, for determining how many continuous derivatives the limit curve of a binary scheme has is as follows.

Let S be the scheme to be analysed and T and U local scheme-valued variables holding the arity and the mask. l_∞ is a function returning the l_∞ norm of a scheme, and square is a function returning the square of a scheme.

```
begin
     lb := -1;
     done := false;
     T := S;
     until  done
     do      if     T has a factor of (1 − zᵃ)/(1 − z)
             then   T := divided difference of T;
                    U := T;
                    while  l∞(U) ≥ U.arity
                    and there is enough memory to hold U²
                    do     U := square(U);
                    od
                    if     l∞(U) < U.arity
                    then   lb := lb + 1;
                    else   done := true;
                    fi
             else   done := true;
             fi
     od
return lb as the number of continuous derivatives.
end
```

If the value of lb returned is -1, that means that the limit curve itself is not proven continuous: if the value of lb is 0, it means that the limit curve is continuous but it does not necessarily have a continuous derivative. If the value is higher, it means that the lb^{th} derivative is continuous, but not the $lb + 1^{th}$.

This procedure was still the state of the art at about the year 2000, but we shall do better using techniques to be found in the next few chapters.

16.7 Exercises

(i)   How many continuous derivatives does the ternary neither scheme $[1,3,5,5,3,1]/6$ have ?

(ii)   Implement the pseudocode from 16.6 above for determining integer lower bounds on Hölder continuity.

16.8 Summary

(i)   Integer lower bounds for continuity can be determined by taking a sequence of divided difference schemes, and checking each for continuity, by testing its difference scheme for contractivity.

(ii)   This involves algebraically dividing the z-transform of the mask by $\frac{1-z^a}{a(1-z)}$, which equals $(1+z)/2$ for binary schemes.

(iii) The bounds achieved in this way can often be tightened by applying the same procedure to higher powers of the scheme.

17. Continuity 4 - Difference Eigenanalysis

Promenade

The next twist of the plot links the two strands encountered so far, using the ideas of the divided difference schemes, expressed in terms of z-transforms, to make the upper bound eigenanalysis dramatically easier.

17.1 Efficient Computation of the Eigencomponents

The original scheme and its divided difference scheme are clearly closely linked, and an interesting question to ask is whether their eigenfactorisations are related in any transparent and useful way.

Yes, they are !

Suppose that V is an eigencolumn with eigenvalue λ, of a scheme S, of arity a, whose mask is M.

Let the divided difference scheme of S be S'.

When S is applied to V, the result is a copy of V, scaled by λ. The first differences of V are therefore also scaled by λ, and so the first divided differences are scaled by $a\lambda$.

Thus the first differences of an eigencolumn of S form an eigencolumn of S', and the corresponding eigenvalue is scaled up by a factor of a.

Note that the unit eigencolumn vanishes in a puff of smoke, because its first differences are all zero. Yes, a column of zeroes is an eigenvector, but it is the trivial one, not to be considered beside the real ones. The number of eigencomponents of the divided difference scheme is therefore one less than the number in the original scheme.

17.1.1 The Kernel

Going from the scheme to its difference scheme removes the unit eigencomponent, and each of the other components has its eigenvalue multiplied by the arity and its (column) eigenvector converted by simple differencing. We can do this as many times as there are σ factors in the generating function of the original scheme.

M. Sabin, *Analysis and Design of Univariate Subdivision Schemes*, Geometry and Computing 6,   103
DOI 10.1007/978-3-642-13648-1_17, © Springer-Verlag Berlin Heidelberg 2010

Working the other way is even more interesting. We can take the eigenanalysis of whatever scheme is left when all of the σ factors have been removed, and progressively work back by taking each eigencomponent, dividing its eigenvalue by the arity and anti-differencing the eigenvector, and then adding a unit eigencomponent to the set.

Whatever is left when all of the σ factors has been removed is of such importance that we give it a name, the **Kernel**[25] of the scheme.

Our upper bounds on continuity can therefore be evaluated with a lot less arithmetic than in chapter 15 above. If the scheme can be factorized into

$$\sigma^k\, K(z)$$

then we take the largest eigenvalue, e, of the matrix of K and divide by a^k. The Hölder continuity is bounded above by

$$-\log_a(|e|/a^k) = -\log_a(|e|) + \log_a(a^k) = k - \log_a(|e|)$$

Of course, this gives only the dyadic[26] bound on continuity, but all of this applies equally to the high arity schemes obtained by raising the kernel to a high power. This makes everything more economical, because if the original scheme has a width of w, the kernel has a width of only $w - (a-1)k$. Both the raising to powers and the eigenanalysis become vastly easier.

17.1.2 Eigenvectors

Once the eigenvalues are determined, it is always possible to get the eigenvectors from the original subdivision matrix. It is also possible to determine them in parallel with the eigenvalues by working back down the chain of divided difference schemes.

We saw above that each eigencolumn of the original scheme gives an eigencolumn of the divided difference scheme by just differencing. Working back requires antidifferencing. This has but one complication, which is that anti-differencing, like antidifferentiation, requires a constant of integration. This has to be determined by solving a single linear equation.

The eigenrows are somewhat easier in that we have to antidifference on the way up the chain and therefore only have to difference on the way back down. The complication takes a different form, deciding what eigenrow is going to apply to each unit eigencomponent as it gets inserted. It turns out

---

[25]Note that the kernel itself is *not* a subdivision scheme. If you try to use it as one it will not converge. However, it has an arity and a sequence of values of some specific length, and so it can be stored in the same shape of computational object as if it were, and it can be operated on in the same ways. Although the entries of each of its stencils do not sum to unity, the sum of the terms in its mask does equal the arity.

[26]in the binary case: for higher arities the successive refinements are by the arity.

that the constants of integration required for the eigencolumns are exactly what is needed, applied in a different way, to give the right eigenrow.

The unit column eigenvector is a polynomial sequence (of degree 0), and anti-differencing any polynomial merely gives a polynomial of degree higher by 1. Thus every $(1 + z)$ factor in the original gives an additional polynomial component, and the dominant non-polynomial eigencomponent in the scheme must come from the kernel. Its eigenvalue is just the dominant eigenvalue of the matrix of the kernel, divided by the arity raised to a power the number of $(1 + z)$ factors.

The viewpoint of this chapter gives us a good argument as to what the 'right' scale is for each row eigenvector (and thence for the columns). The argument is simple: it is possible to scale the row eigenvectors corresponding to polynomial columns so that applying them to the original polygon gives the values of derivatives at points on the limit curves. This happens automatically if we introduce the unit eigenrows at such a scale that their components sum to 1. Subsequent differencing turns them into exactly the right stencils.

17.2 Examples

17.2.1 Cubic B-spline

The mask $(1 + 4z + 6z^2 + 4z^3 + z^4)/8$ is $2((1 + z)/2)^4$, and so we can go back to the scheme whose mask is [2] in just four steps. The matrix of this has an eigenvalue of 2, and its row and column eigenvectors are both symmetric. (It is this component which gives rise to the discontinuity.)

After one step back down the chain we have an eigenvalue of 1, given by dividing 2 by the arity, and another eigenvalue of 1, added to the list in the usual way. After two steps we have eigenvalues of $1/2,1/2$ by dividing these by the arity plus a unit eigenvalue. After three the list is $1/4,1/4,1/2,1$, and after four $1/8,1/8,1/4,1/2,1$.

No solutions of quadratic equations at all were needed, and a very visible structure emerges, with reasons, for the last four components.

17.2.2 Four point

Consider the four point scheme again. Its kernel is $(-1 + 4z - z^2)$ and the kernel subdivision matrix is

$$
\begin{bmatrix}
\ddots & & & \\
& -1 & -1 & \\
& & 4 & \\
& & -1 & -1 \\
& & & \ddots
\end{bmatrix}
$$

The eigenanalysis is now trivial. The antisymmetric component has only the trailer, -1, the symmetric component is block lower triangular and its eigenvalues are 4 and -1. When we take these through two successive multiplications by $(1 + z)/2$, the dominant eigenvalue becomes $4/2^2 = 1$ which is the same as the unit eigenvalue naturally appearing at that stage. There is a coupling between the two components, as we shall see when we look at eigenvectors below, and this causes a Jordan block.

17.3 Dual Schemes

Dual schemes are handled in exactly the same way. Each taking of a divided difference scheme switches either from primal to dual or the reverse, and so both primals and duals are intimately involved in any scheme.

17.4 Higher Arity Schemes

For higher arity schemes we have the two complications noted in the previous chapter, that the equivalent of the $(1+z)/2$ found in the binary case becomes $(1 + z + \ldots + z^{a-1})/a$.

Then we find that the eigenanalysis can be performed at $a-1$ mark points spread along a span of the polygon.

17.5 A Special Case

When we raise the kernel of a B-spline to a higher power we get again a single entry in its mask. When this is converted to a matrix we find the interesting effect that the submatrices corresponding to diagonals other than through that single entry are of size 0×0, which have no eigencomponents.

For example, when the kernel is [2] of a binary scheme, the kernel of its quaternary square is [4], giving the matrix

$$\begin{bmatrix} 4 & \\ 0 & 0 \\ 0 & 0 \\ 0 & 0 \\ 0 & 4 \end{bmatrix}$$

The process of integrating back up, for those submatrices which do not contain the '4', gives schemes with only polynomial eigenvectors. There is no discontinuity at all. This fits exactly what we know about B-splines, which are indeed fully continuous at all points in the interior of the spans.

17.6 Exercises

(i)   What is σ for a ternary scheme ?

(ii)   What is the kernel of the ternary neither scheme [1,3,5,5,3,1]/6, and how many σ factors does it have ?

(iii)   Write a routine to determine k and $K(z)$ for a given scheme.

(iv)   Write a routine for determining the Hölder continuity upper bound from k and $K(z)$.

(v)   Write a routine to determine the appropriately scaled row eigenvectors to give the limit points and whatever derivatives exist at the mark points.

17.7 Summary

(i)   Factorisation of the symbol of the scheme into factors of σ and a kernel can reduce the size of matrix to be analysed dramatically.

(ii)   The eigenvalues obtained from analysis of the kernel are then scaled down by a factor of the arity for each smoothing factor. The eigenvectors can also be obtained progressively as the σ factors are reapplied.

(iii)   The column eigenvectors added in this way are all polynomial and so only the kernel eigencomponents contribute to discontinuity.

(iv)   The row eigenvectors give stencils which, applied to the original polygon, give points and derivatives on the limit curve.

18. Continuity 5 - the Joint Spectral Radius

Promenade

We have seen how a lower bound on the continuity of the limit curve can be determined by z-transform analysis, and an upper bound by eigenanalysis.

The insights of z-transforms enable us to compute the eigenproperties very efficiently.

Yet another twist shows that both of these two computations can be understood in terms of a standard property of a pair of matrices, their **Joint Spectral Radius**.

18.1 The Joint Spectral Radius Approach

In the eigenanalysis chapter we used the support analysis to tell us how many control points influenced a neighbourhood of a point of the limit curve, and thus how large the matrix needed to be on which to carry out the eigenanalysis.

However, this told us only the continuity exactly at one point of the limit curve.

We can also ask the support analysis how many points influence one span of the limit curve, the piece corresponding to one edge of the control polygon. This turns out to be one fewer. Call it m. The value will be 4 for the cubic B-spline.

After one refinement, the pieces of limit curve are just half as long, and there are two pieces, each dependent on m new points.

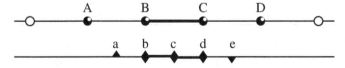

The left hand half depends on the new points a to d, the right hand half on b to e, all of which depend only on A to D.

M. Sabin, *Analysis and Design of Univariate Subdivision Schemes*, Geometry and Computing 6,   109
DOI 10.1007/978-3-642-13648-1_18, © Springer-Verlag Berlin Heidelberg 2010

Thus there are two $m \times m$ sub-matrices within the overall subdivision matrix, sharing the same non-zero columns. Their rows overlap by $m - 1$, because the sets of new control points influencing the two new pieces are adjacent consecutive sequences.

$$
8 \begin{bmatrix} a \\ b \\ c \\ d \end{bmatrix} = \begin{bmatrix} \ddots & & & \\ 1 & 6 & 1 & & \\ 4 & 4 & 0 & 0 \\ 1 & 6 & 1 & 0 \\ 0 & 4 & 4 & 0 \\ 0 & 1 & 6 & 1 \\ 0 & 0 & 4 & 4 \\ 0 & 0 & 1 & 6 & 1 \\ & & & & \ddots \end{bmatrix} \begin{bmatrix} A \\ B \\ C \\ D \end{bmatrix}
$$

and

$$
8 \begin{bmatrix} b \\ c \\ d \\ e \end{bmatrix} = \begin{bmatrix} \ddots & & & \\ 1 & 6 & 1 & & \\ 4 & 4 & 0 & 0 \\ 1 & 6 & 1 & 0 \\ 0 & 4 & 4 & 0 \\ 0 & 1 & 6 & 1 \\ 0 & 0 & 4 & 4 \\ 0 & 0 & 1 & 6 & 1 \\ & & & & \ddots \end{bmatrix} \begin{bmatrix} A \\ B \\ C \\ D \end{bmatrix}
$$

The upper such matrix tells how the original m control points influence those of the left-hand subpiece, the lower submatrix how they influence the right hand subpiece. Call the two submatrices L and R.

At the next iteration there are four pieces, which are given by LL, RL, LR and RR times the original piece of polygon. Note that these four matrices are still $m \times m$ in size and they all overlap, being successive m-line pieces of the matrix of the arity 4 scheme given by applying the original scheme twice.

After three refinements there are eight pieces, given by sequences three-long of L and R.

L				R			
LL		RL		LR		RR	
LLL	RLL	LRL	RRL	LLR	RLR	LRR	RRR

After many refinements, each short segment of limit curve is given by some sequence of upper and lower choices in the instances of the matrix in the high power which corresponds to making many refinements. If we want to consider what might happen anywhere in the limit curve we need to consider all such sequences.

This is what the joint spectral radius analysis does. The joint spectral radius of the two matrices L and R, is defined as the limit, as n tends to ∞, of the value of the nth root of the largest dominant eigenvalue of any of the matrices formed by taking all possible product sequences of length n of L and R.

The key theorem is that upper bounds, given by norms, and lower bounds, given by actually taking eigenvalues as per the definition of joint spectral radius, do in fact converge to the same value.

An upper bound on this is given for finite n by the maximum matrix norm taken over all the matrices given by such sequences of length n. This bound converges as n increases. It is worth noting that because these matrices are just the submatrices of the high arity one, they overlap, each sharing all of its rows bar the last with the previous and all rows bar the first with the next. This means that to determine the highest l_∞ norm of all the submatrices it is only necessary to scan once through all the rows of the long thin rectangular matrix, not through the rows of each submatrix.

A lower bound is given by the largest non-polynomial eigenvalue, and clearly this lower bound also converges as n increases.

Unfortunately there is no simple method of computing directly the joint spectral radius for two given general matrices. The best we can do is to compute the upper and lower bounds for larger and larger values of n and watch them converge.

These computations are exactly the computations we would have carried out using the eigenanalysis and z-transform approaches, using higher and higher powers of the scheme.

There are, however, some valuable tricks to play to ease these calculations. Again they depend on factorising the scheme into a kernel and σ factors.

18.2 The Continuity Argument

If the dth divided difference scheme S_d of some power of a scheme S has an l_∞ norm less than the arity of that power, then the difference scheme of the $(d-1)^{th}$ divided difference has an l_∞ norm less than one, and so the $(d-1)^{th}$ divided difference scheme is contractive, and therefore converges. The limit curve of the original scheme has continuity of the $(d-1)^{th}$ derivative.

If the dth divided difference scheme S_d of some power of a scheme S has a dominant eigenvalue greater than or equal to the arity of that power, then the $(d-1)^{th}$ divided difference scheme has an eigenvalue greater than or equal to 1, which will dominate over any polynomial eigencomponents, and so the limit curve of the original scheme has a discontinuity of $(d - 1)^{th}$ derivative.

Note that the kernel itself always has an l_∞ norm greater than 1, because the sum of the entries in the mask is equal to the arity a. It is also equal to the sum of the stencil sums and there are a stencils. If the average stencil sum is 1, and they are not all equal to 1 (when the mask would be divisible by $(1 - z^a)/(1 - z)$, which the kernel is not) at least one must be greater than 1.

The subtlety of the joint spectral radius results is that the fact that both the highest eigenvalue and the lowest l_∞ norm converge to a well defined joint spectral radius value means that the l_∞ norm of any finite power of the kernel gives an upper bound on the eigenvalues of all powers of the kernel, and therefore a lower bound on any constructible discontinuity. We do not have to prove that the l_∞ norm shrinks by the arity every time we add an extra $(1 - z^a)/a(1 - z)$ factor.

18.3 A Procedure for Determining Hölder Continuity

Given a scheme S of arity a

```
    begin
        factorise S(z) into σₐᵏK(z)
        raise K(z) to a high power Kⁿ(z)
        determine the l∞ norm l of Kⁿ(z)
            and the rows with the largest sum of absolute values.
        for each submatrix M containing any of these rows
        do       determine the dominant eigenvalue of M
                    recording the largest, e
        od
    end
```

An upper bound on the Hölder continuity of S is $k - \log_a(|e|)/n$ and a lower bound $k - \log_a(|l|)/n$, except that if either bound is an integer i the Hölder bound is $i - 1, 1$ rather than i.

If even tighter bounds are required, then a search further down the binary tree of submatrices can be undertaken, starting from those matrices whose norm is greater than the largest eigenvalue so far, e. A 2^n-ary tree may in fact be more efficient, because each matrix with norm $l > e$ need only be multiplied by those matrices with norm greater than e^2/l.

Also starting by getting a value of e from the original scheme, its square and its cube may be a good strategy for being as selective as possible when choosing submatrices to calculate eigenvalues for later. Because of symmetries, A, AB and AAB will be enough to look at for this purpose.

Be warned that there are schemes with coefficients close to B-splines where almost all the submatrices in the tree have almost exactly the same eigenvalues. For these the above algorithm may be extremely slow.

18.4 Summary

(i)   The joint spectral radius of the kernel leads in principle to the exact Hölder continuity of the scheme, but unfortunately we do not have an efficient algorithm for going directly to its exact value.

(ii) The computations giving bounds on the joint spectral radius are essentially the computations that we would have carried out with the previous approaches, of checking contractivity by norms and of eigenanalysis, but this result unifies the two approaches, once thought competing, in a very elegant way.

19. What Converges ?

Promenade

There is an apparent contradiction which has been swept under the carpet in the previous few chapters.

We have gaily elaborated very plausibly on how to compute the continuity levels of schemes with high continuity, but the definition of the limit curve has been as the limit of a sequence of polygons. Now each polygon has a piecewise constant first derivative, and its second derivative is a sequence of Dirac delta functions. As we refine, the sequence of delta functions becomes denser and denser until in the limit there is one at every dyadic value of abscissa. Although the dyadics are dense in the reals, there are lots of reals which are not dyadics, and so the second derivative of the limit of any sequence of polygons is going to be something like the Dedekind function, with delta functions interspersed by zeroes at an extremely dense scale. How can we talk about schemes having a continuity of C^2 or higher, which have this kind of structure ?

19.1 A More Appropriate Description

There is a more appropriate wording that can be used.

If a scheme has a C_k limit, then the polygons converge towards that limit.

This is not at all objectionable, because we can make examples of polygons converging towards even a C_∞ function (such as $\sin(x)$) merely by joining samples evaluated from the desired function. As the samples are taken more and more densely, the polygons converge. The order of convergence is quadratic if the function is C_1 or more, linear if it is only C_0.

M. Sabin, *Analysis and Design of Univariate Subdivision Schemes*, Geometry and Computing 6,   115
DOI 10.1007/978-3-642-13648-1_19, © Springer-Verlag Berlin Heidelberg 2010

19.2 A More Appropriate Definition

However, defining the limit function as the limit of the polygons is prob-
lematic, and it can be avoided, merely by defining it instead as the limit
of a sequence of C_k functions. The question then is *"what sequence of what
functions?"*.

This is very simple. We can define it as the limit of a sequence of any well-
behaved curve[27] defined by the vertices of the polygons as control points. For
example, we could define it as the limit of

$$f_j(x) = \Sigma_i f_{ij}(\sin(2^j x)/2^j x)$$

Every curve in the sequence would be C_∞, but the limit would have
divergent derivatives at the places where the analysis in previous chapters
predicted it. This behaviour is not regarded as in any way abnormal. For
example, $\tan^{-1}(nx)$ is C_∞ everywhere for all finite n, but the limit as $n \to \infty$
has a discontinuity at $x = 0$ where the first derivative diverges.

However, there is a better choice. If we believe the limit function to be
C_k, then we choose as our sequence of functions the B-splines of degree $k+1$.
This is actually the assumption which has been made implicitly in the use
of difference schemes above. If, for example, a scheme has a convergent second difference scheme, then we can think of this second difference scheme as
having a polygon, so that the second derivative varies linearly. If the second
derivatives vary linearly in each span, then the first derivatives vary quadratically and the actual values cubically, and so, because the support is finite,
the implicit approximant is a sequence of cubic B-splines[28].

This definition is also justified by the view taken of the cascade algorithm
at the end of section 12.1.

19.3 Example

This view can be illustrated by seeing how the cubic B-spline approximants
to the basis function of the four-point scheme converge.

---

[27]'well behaved' here probably means that there is a well-defined enclosure with
not too large an expansion factor.
[28]That the pieces are cubic is evident. Because the pieces join with continuity of
second derivative the pieces form a spline. Because the support is bounded, they
form a B-spline.

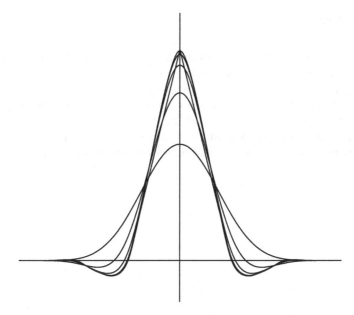

An even more interesting picture emerges if we plot the second derivatives of the cubic B-spline approximations at successive steps. In the next figure each line is a plot of the B-spline second derivative at a dyadic place near to $t = 1/3$. The steps are highlighted by vertical bars.

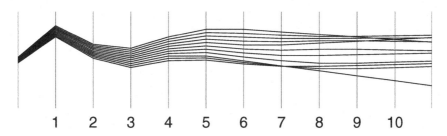

The first to start diverging (the third from bottom at the left) is at a multiple of $1/32$ and it starts diverging at the 5th bar. The next is at the top, at a multiple of $1/64$, diverging from the 6th bar. In between, the multiple of $1/128$ at 5 from the top diverges from the 7th.

The larger the denominator of the dyadic point, the later its divergence starts, and the slower the divergence because by then the fourth differences (which are what drive the divergence) are relatively small. If the figure were continued a long way to the right, all of these plots would diverge, but a plot exactly at $t = 1/3$ would remain convergent.

19.4 Summary

(i)   An apparent contradiction evaporates if we regard the limit curve as being defined, not by the sequence of polygons, but by a sequence of B-splines.

(ii)   In the case of B-spline schemes the convergence is immediate, but for more interesting ones it is typically quadratic.

(iii) Following this for a particular example helps to illustrate how the convergence behaviour can be different for different abscissae.

20. Reproduction of Polynomials

Promenade

What happens when the control points of the initial polygon have values following some pattern ?

A classical question is to consider the case when the control values are samples taken from some polynomial, and to ask how the limit function is related to that polynomial.

There are actually three questions with three different answers.

i   If the control values are sampled at equal intervals from a polynomial of degree d, up to what value of d will the limit function be a polynomial ? We call the maximum value of d the **spanning degree** or **generation degree** of the scheme.

ii If the control values are sampled at equal intervals from a polynomial of degree d, up to what value of d will the limit function interpolate the sample points ? We call the maximum value of d the **interpolating degree** or the **quasi-interpolating degree** of the scheme.

iii If the control values are sampled at equal intervals from a polynomial of degree d, up to what value of d will the limit function reproduce that polynomial ? We call the maximum value of d the **reproduction degree** of the scheme. Polynomials of degree not exceeding this value form the **precision set** of the scheme.

The key fact on which we build a theory is that if (and only if) polygon vertices are samples at equal intervals from a polynomial of degree at most d, then the $d+1^{th}$ differences (and all higher differences) of the polygon are zero.

i.e. if P_0 is such a polygon, then $(1 - z)^{d+1} P_0(z) = 0$

20.1 Generation Degree

We consider first binary schemes: primal and dual need not be distinguished.
If the symbol has $d+1$ factors of $(1+z)/2$ then

$$
\begin{aligned}
(1-z)^{d+1}P_1(z) &= (1-z)^{d+1}S(z)P_0(z^2) \\
&= (1-z)^{d+1}((1+z)^{d+1}/2^{d+1})K(z)P_0(z^2) \\
&= (1/2^{d+1})K(z)(1-z)^{d+1}(1+z)^{d+1}P_0(z^2) \\
&= (1/2^{d+1})K(z)(1-z^2)^{d+1}P_0(z^2) \\
&= (1/2^{d+1})K(z)*0 \\
&= 0
\end{aligned}
$$

so that if the original polygon has vertices on a polynomial of degree d (whose $d+1^{th}$ differences are zero), the $d+1^{th}$ differences of the refined polygon are also zero and thus the refined polygon vertices also lie on a polynomial of the same degree.

If the symbol has fewer factors, e say, of $(1+z)/2$ then

$$
(1-z)^{e+1}P_1(z) = K(z)(1-z^2)^{e+1}P_0(z^2)
$$

which would be zero only if the degree of the original polynomial were actually only e.

20.1.1 Higher arities

The same result holds for higher arities. The smoothing factor is $\frac{1-z^a}{a(1-z)}$ and so exactly the same proof logic applies.

20.1.2 Local polynomial structure

Remember that a subdivision limit curve consists of pieces, each of which depends only on a finite number, w, of original control points. If those control points lie on a polynomial, then that piece of the limit curve will be polynomial or not, depending on the degree, irrespective of the other control points.

However, a polynomial of degree $w-1$ can always be fitted, however many such control points there are, and if $w-1$ is less than or equal to the spanning degree, the limit curve will have polynomial pieces.

Consider first the binary case. Let the scheme be of the form $\sigma^n K$ where K is the kernel of k entries.

The total width w of the mask is $k+n$ because each convolution with σ adds one to the width.

The number of control points influencing one span is $w - 1$. We can always find a polynomial of degree at most $w - 2$ interpolating all of these points. That polynomial will be spanned if its degree is less than or equal to $n - 1$ and so we have the condition

$$w - 2 = k + n - 2 \leq n - 1$$

or

$$k \leq 1$$

The only solution is $k = 1$, so that the only binary schemes with piecewise polynomial limit curves are the B-splines.

Unfortunately this result is not quite true. In order to understand the exceptions we shall need the concept of a two-slanted sampling matrix which will be introduced in chapter 21, and so this whole issue will be taken up in chapter 22. Essentially factors other than $(1 + z)/2$ can have the effect of increasing the effective degree. The theory here does apply to **normalised** schemes, where those other factors have been removed.

20.1.3 Higher arities

Now consider higher arities. Again the mask has $k + n(a - 1)$ entries and the width w of the basis function is $(k + n(a - 1) - 1)/(a - 1)$. If the width is not an exact integer there will be pieces of limit curve influenced by $c = \lfloor \frac{w-1}{a-1} \rfloor$ control points and other pieces influenced by $\lceil \frac{w-1}{a-1} \rceil$.

The former piece can have control points lying on a polynomial of degree $c - 1$ and this will be spanned if $c - 1 \leq n - 1$.

$$\lfloor \frac{k + n(a - 1) - 1}{a - 1} \rfloor - 1 \leq n - 1$$

$$\lfloor \frac{k - 1 + n(a - 1)}{a - 1} \rfloor \leq n$$

$$\lfloor \frac{k - 1}{a - 1} \rfloor + n \leq n$$

$$\lfloor \frac{k - 1}{a - 1} \rfloor \leq 0$$

$$k - 1 < a - 1$$

$$k < a$$

The latter piece requires $k \leq 0$ which will never be satisfied.

The ternary neither example (Example 4) shows this behaviour. The mask is

$$[1, 3, 5, 5, 3, 1]/6 = [1, 1, 1]^2 [1, 1]$$

so that $a = 3, k = 2, n = 2$.

$k < a$ and therefore there can be polynomial pieces of degree $n - 1 = 1$.

Looking at the support width, we see that at each refinement just half of the remaining amount of parameter line is filled with straight line pieces. Thus as refinement steps proceed, the amount remaining is halved at each step: the limit curve consists entirely of pieces of straight line.

With higher arity still the kernel width can be higher. The fraction of each span with the lower number of support points is $(k - 1)/(a - 1)$, and this fraction is filled at the first step of refinement by a polynomial piece. The remainder is a fraction $r = (a - k)/(a - 1)$ which is less than 1. It has that same fraction as the original filled at the next step by one or more polynomial pieces, and thus the amount remaining unfilled by polynomial pieces after m steps is r^m which converges to zero. Thus the entire limit curve consists of polynomial pieces (of degree $n - 1$), but an unbounded number of them.

20.2 Interpolating Degree

20.2.1 Primal binary schemes

Consider the primal binary case first.

The unit eigenvector gives the stencil for a point on the limit curve corresponding to an original control point. Express the symmetric form of its stencil as a polynomial in $\delta^2 = (1 - z)^2$.

$$\alpha_0 + \alpha_1 \delta^2 + \alpha_2 \delta^4 \dots$$
$$\sum_{i \in 0..\lfloor n/2 \rfloor} \alpha_i \delta^{2i}$$

The value of α_0 is always 1. The interpolating degree is determined by the subscript, i, of the first non-zero α_i.

$$p(z) = \alpha_0 P(z) + \alpha_1 (1 - z)^2 P(z) + \alpha_2 (1 - z)^4 P(z) + \dots$$
$$p(z) - P(z) = \alpha_1 (1 - z)^2 P(z) + \alpha_2 (1 - z)^4 P(z) + \dots$$

where P is the sequence of original control points and p is the sequence of corresponding limit curve points.

If P is sampled from a polynomial of degree d, then the right hand end of the series is truncated because a sufficiently high power of δ (greater than d) annihilates the term. If sufficiently many α_i are zero, the left hand end of the series is also zero, so that $p(z) = P(z)$, which is the interpolation condition.

It is equally possible to take the stencil for the v-vertices, and express that as a polynomial in δ^2. There is then a direct argument that, if sufficient α_i are zero, then for a polynomial of sufficiently low degree, at every stage the new v-vertices coincide with the old ones, and thence the limit curve must do so also.

20.2.2 Dual binary schemes

When the scheme is not primal, the scheme maps each vertex into an edge. However, the unit row eigenvector as a stencil maps edges into vertices, and so the product of the mask with the unit row eigenvector maps vertices into vertices.

This product is now at a higher density, and so we need to select just those terms which are at the original density (just as we selected terms in the v-stencil in the primal case).

In fact this product can be used in the primal case also, but the individual factors are shorter and therefore easier to handle.

20.2.3 Higher arities

Even at higher arities the formula of multiplying the mask by the appropriate unit row eigenvector and selecting just the vertices corresponding to the original vertices, then expressing that stencil as a polynomial in δ^2 works. Where the scheme is primal the short cut of just looking at the v-vertex stencil also works.

20.3 Reproduction Degree

This is the classical question, because it tells how densely the points on some other curve need to be sampled in order to give a desired accuracy of fit between the original curve and the approximant. If the degree of reproduction is d, then the order of approximation is $O(h^{d+1})$, and so every doubling of the density of the sampling gives a reduction in the error by a factor of 2^{d+1}.

Note that this doubling of the density is not the refinement of the subdivision step, but a doubling of the density of the original polygon. A doubling of the amount of work done in collecting data.

We still need to be concerned about it.

If a given polynomial is to be reproduced, it must be within the span of the scheme, and it must also be interpolated. We can therefore see without any sophisticated argument that the degree of reproduction is just the lower of the spanning degree and the interpolation degree.

20.4 Exercises

(i)   Check the reproduction degree of the four-point scheme [-1,0,9,16,9,0,-1]/16.

(ii) Write a routine to convolve the mask with the unit row eigenvector (multiply their z-transforms) and express the even entries in this product as a polynomial in δ^2. Then count the zeroes near the beginning.

20.5 Summary

(i)   The maximum degree of polynomial, the spanning degree, which a scheme can have in its limit curve is $n - 1$ where n is the number of $(1 + z)/2$ factors in the symbol.

(ii) If the width of the kernel is 1, then the limit curve will always consist of pieces of this degree.

(iii) If the width of the kernel is less than the arity, the limit curve will consist of a fractal assembly of pieces of this degree.

(iv) The maximum degree of polynomial with the property that if the control points all lie on it they will be interpolated by the limit curve, the interpolating degree, can be determined by looking at the limit stencil (a subsequence of the terms in the convolution of the mask and the unit row eigenvector) as a polynomial in $\delta = (1 - z)^2$.

(v)   The classical degree of reproduction, which gives the approximation order of the scheme, is just the lesser of the spanning degree and the interpolating degree.

21. Artifacts

Promenade

We have seen the use of two particular data patterns, **cardinal data** where only one control point has a unit value and all others are zero, and **polynomial data** where all control points have values lying on some polynomial.

Cardinal data led us to the basis function; polynomial data led to global approximation properties of the limit curve.

We have also analysed what continuity the limit curve has. But continuity is essentially a local property. We now look at what happens with yet another data pattern, that when all control points have values lying on a sinusoid, and this tells us about structures intermediate in scale between the two.

21.1 Artifacts

When a designer is checking the fairness of a curve one of the tools which can be used is the 'curvature plot'. This is a graph of the curvature against arc length.

It is quite typical for curves which are defined in terms of a control polygon, such as B-splines and subdivision curves, to exhibit features in their curvature plots with a spatial frequency too high to be justified by the data.

For example, the quadratic B-spline fitted to four points per cycle, has a curvature plot where the maximum and minimum curvatures have a ratio greater than 2:1.

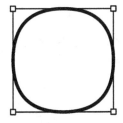 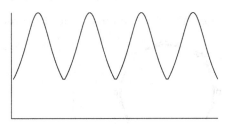

The cubic B-spline has a ratio of just under 2:1.

M. Sabin, *Analysis and Design of Univariate Subdivision Schemes*, Geometry and Computing 6,   125
DOI 10.1007/978-3-642-13648-1_21, © Springer-Verlag Berlin Heidelberg 2010

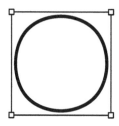 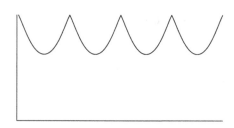

In both of these examples there is a maximum of curvature near each control point and a minimum in between. There is no systematic movement of the control points which can remove this variation. If proof of this is required, we can appeal to the Shannon sampling theorem. Although this was originally conceived in the temporal domain it applies equally to the spatial.

The ripples in the curvature plot are essentially created by the definition of the limit curve in terms of the vertices. They are **artifacts** of that construction.

This is not due to subdivision, but is a property of the splines themselves. However, we can use the subdivision construction to analyse just how bad the effect is, not just for splines, but for any curve which can be constructed by subdivision.

Because the ripples have the spatial frequency that they do, we can perceive that most of the artifact is caused by the first step of subdivision. Beyond that step the artifacts introduced then are just part of the definition. It is also evident from a few experiments that the ripple amplitude is strongly influenced by the spatial frequency of the original sampling.

For example, if we use eight points per cycle, the curvature plot of the quadratic is significantly improved,

 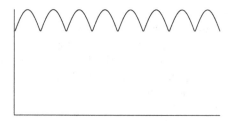

and for the cubic the improvement is even more marked.

 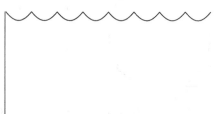

We therefore look first at what happens during a single step, with data at just a single spatial frequency.

21.2 Factorising the Subdivision Matrix

The subdivision matrix of a binary uniform scheme has each column a copy of the mask, shifted down two rows for each step to the right. Every row (stencil) sums to 1, but the mask entries sum to 2.

We can reexpress this matrix as the product of two matrices. The first is a circulant matrix with every row[29] and every column a copy of the mask divided by 2. The second is a 'diagonal' matrix with the value 2 on the diagonal of slope -2. We call the latter the **sampling matrix** and the former the **smoothing matrix**.

For example, the scheme whose mask is $[1, 2, 1]/2$ can be written in matrix form as

$$
\begin{bmatrix}
\ddots & & & & \\
2 & & & & \\
1 & 1 & & & \\
& 2 & & & \\
& 1 & 1 & & \\
& & 2 & & \\
& & & \ddots &
\end{bmatrix}/2 =
\begin{bmatrix}
\ddots & & & & \\
2 & 1 & & & \\
1 & 2 & 1 & & \\
& 1 & 2 & 1 & \\
& & 1 & 2 & 1 \\
& & & 1 & 2 \\
& & & & \ddots
\end{bmatrix}/4
\begin{bmatrix}
\ddots & & & & \\
2 & & & & \\
0 & 0 & & & \\
& 2 & & & \\
& 0 & 0 & & \\
& & 2 & & \\
& & & \ddots &
\end{bmatrix}
$$

Essentially, alternate columns in the smoothing matrix are multiplied by the empty rows of the sampling matrix.

21.2.1 Effect of the sampling matrix

This inserts zero values at the e-vertex positions and doubles whatever values were originally present to give v-vertices.

If the original polygon is sampled from a sinusoid of spatial frequency $\omega = 1/m$ where m is the number of polygon points per complete cycle, so that $P_0[j] = cos(2\pi j\omega)$, $j \in \mathbb{Z}$, the result of multiplying by the sampling matrix can be expressed as

$$cos(2\pi j\omega)(1 + cos(2\pi j)), j \in \mathbb{Z}/2$$

We can view this as the sum of a **signal component**, $cos(2\pi j\omega)$, and an **artifact component**[30], $cos(2\pi j\omega)cos(2\pi j)$

---

[29]Strictly speaking the rows are the reverse of the mask, but because we deal with palindromic masks we ignore that here.

[30]This can further be expressed in terms of two **sideband** components, $cos(2\pi(\omega \pm 1)j)$, but for this analysis it is convenient not to do so.

The two components each have unit amplitude.

21.2.2 Effect of the smoothing matrix

The smoothing matrix can now be viewed as a filter which will have different effects on the two components. We attack the analysis of its effects by
doing a further factorisation, exploiting the fact that the z transform can be
regarded as a representation of the smoothing matrix as a polynomial in the
unit subdiagonal. Factorisation of a circulant matrix is exactly equivalent to
factorisation of the generating function as a polynomial.

This factorisation is made unique by the convention that every factor
must have rows which sum to unity. Also, because the mask is palindromic,
we use the form of the symbol divided by such a power of $\sqrt{z}$ as will make
the coefficients of positive and negative powers of $\sqrt{z}$ equal. Thus $\sigma = (1 + z)/2\sqrt{z}$.

We first pull out as many factors of $\sigma^2 = \left(\frac{(1+z)^2}{4z}\right)$ as we can.

Each such factor has the effect of scaling the signal and artifact components by factors determined as follows.

Three consecutive original signal components after sampling are

$$cos(2\pi\omega(j-1/2),$$
$$cos(2\pi j\omega)$$
$$\text{and } cos(2\pi\omega(j+1/2),$$

so that when the signal is multiplied by σ^2 we get

$$(cos(2\pi(j-1/2)\omega) + 2cos(2\pi j\omega) + cos(2\pi(j+1/2)\omega))/4$$
$$=(2cos(2\pi j\omega)cos(2\pi\omega/2) + 2cos(2\pi j\omega))/4$$
$$=2cos(2\pi j\omega)(1 + cos(\pi\omega))/4$$
$$=2cos(2\pi j\omega)(2cos^2(\pi\omega/2))/4$$
$$=cos(2\pi j\omega)cos^2(\pi\omega/2)$$

and thus every sample is multiplied by $cos^2(\pi\omega/2)$.

Three consecutive artifact components after sampling are

$$cos(2\pi(j-1/2)\omega)cos(2\pi(j-1/2)),$$
$$cos(2\pi j\omega)cos(2\pi j)$$
$$\text{and } cos(2\pi(j+1/2)\omega)cos(2\pi(j+1/2)),$$

so that when the artifact is multiplied by σ^2 we get

$$\left(\begin{array}{c} cos(2\pi(j-1/2)\omega)cos(2\pi(j-1/2)) \\ +2cos(2\pi j\omega)cos(2\pi j) \\ +cos(2\pi(j+1/2)\omega)cos(2\pi(j+1/2)) \end{array} \right) /4$$

$$=(-2cos(2\pi j\omega)cos(2\pi j)cos(2\pi\omega/2) + 2cos(2\pi j\omega)cos(2\pi j))/4$$

$$=2cos(2\pi j\omega)cos(2\pi j)(1 - cos(\pi\omega))/4$$

$$=2cos(2\pi j\omega)cos(2\pi j)(2sin^2(\pi\omega/2))/4$$

$$=cos(2\pi j\omega)cos(2\pi j)sin^2(\pi\omega/2)$$

so that every sample is multiplied by $sin^2(\pi\omega/2)$.

Each of the σ^2 factors in the mask multiplies the amplitudes of the signal and artifact components by these factors.

Primal schemes have an even number of σ factors in the mask, and so all that is left is the kernel of the scheme.

21.2.3 Effect of the kernel

The kernel by definition has no further factors of σ, but it can be expressed as a polynomial in σ. In fact, because the kernel of a binary scheme always has an odd number of entries, its symmetric form can be expressed as a polynomial in σ^2.

Each term in that polynomial multiplies the signal and artifact by an appropriate power of $cos^2(\pi\omega/2)$ or $sin^2(\pi\omega/2)$ respectively, and the resulting contributions can be added together again.

We thus see that the effect of one step of the complete scheme is computed by expressing the mask as a polynomial in σ^2, and then evaluating that polynomial as a function of $cos^2(\pi\omega/2)$ for the signal and $sin^2(\pi\omega/2)$ for the artifact.

21.3 Effect on the Limit Curve

This determines the amount of artifact present in the polygon after the first step of subdivision. What we are really interested in is the amount of artifact present in the limit curve. Once the sampling and smoothing process has been done, we have a denser polygon. We can then apply the unit row eigenvector to this (in the form of a circulant matrix) to determine limit points at this density. It is easiest to carry out the arithmetic by combining the circulant unit eigenrow matrix with the smoothing matrix. That combination can easily be done in the z-transform, where we are merely looking for $(1+z)/2$ factors. We look for them in the eigenrow, too, and then multiply the quotient into the kernel to give a polynomial which is expressed in terms of powers of σ^2.

21.4 Dual Schemes

These have an odd number of σ factors in the mask. However, the unit row eigenvector also has an odd number of σ factors, and so the product has an even number, and so we can determine the amount of artifact in the limit curve in exactly the same way.

The net result for both primal and dual schemes is that if we take the product of the symmetric mask symbol and the symmetric unit eigenrow symbol and express it as a polynomial in σ^2, then the artifact amplitude is given by substituting $\sin^2(\pi\omega/2)$ for σ^2 in that polynomial, and the signal amplitude by subsituting $\cos^2(\pi\omega/2)$.

21.5 Examples

The cubic spline scheme has mask $2\sigma^4$ and unit eigenrow $(2 + 4\sigma^2)/6$. The artifact is therefore given by $2\sin^4(\pi\omega/2)(1 + 2\sin^2(\pi\omega/2))/3$. Near $\omega = 0$ (the case for very dense sampling) this is close to $\pi^4\omega^4/24$.

The quadratic spline scheme has mask $2\sigma^3$ and unit eigenrow σ. The artifact is therefore given by $2\sin^4(\pi\omega/2)$. Near $\omega = 0$ this is close to $\pi^4\omega^4/8$.

The four-point scheme has mask $(6 - 4\sigma^2)\sigma^4$ and unit eigenrow 1. The artifact is therefore given by $2(3 - 2\sin^2(\pi\omega/2))\sin^4(\pi\omega/2)$. Near $\omega = 0$ this is close to $3\pi^4\omega^4/8$.

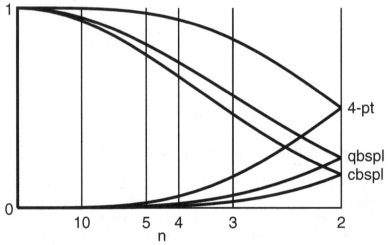

21.6 Higher Arities

For ternary schemes which are both primal and dual there is a short cut. There are two mark points, and we can determine the unit row eigenvector for each of them, thus giving the values at the limit points corresponding to both the vertices and the mid-edges.

For example, if we take the ternary quadratic B-spline, whose mask is [1,3,6,7,6,3,1], the submatrices for the mid-edge and vertex mark points are

$$\begin{bmatrix} 1 & 7 & 1 \\ & 6 & 3 \\ & 3 & 6 \\ & 1 & 7 & 1 \end{bmatrix} \quad \text{and} \quad \begin{bmatrix} 3 & 6 \\ 1 & 7 & 1 \\ & 6 & 3 \end{bmatrix}$$

and their unit row eigenvectors are $[4,4]/8$ and $[1,6,1]/8$

Half the difference of the two eigenrows is a binary mask for the artifact $[1,-4,6,-4,1]/16$, and half the sum a binary mask for the signal $[1,4,6,4,1]/16$. Multiplying the original polygon by these therefore gives the required measures.

It is not too surprising that the ternary and binary quadratic B-splines have the same amounts of signal and artifact. They are the same curve, depending in the same way on the original polygon.

This approach to artifact analysis, while useful for low arities, runs out of steam above ternary. It is certainly possible to determine the limit points corresponding to original vertices and mid-spans, and to create measures of artifact and signal, but this ignores the fact that a quaternary scheme can introduce at the first step artifacts of twice the sampling frequency.

21.7 Exercises

(i)   Determine the artifact amplitude of the ternary neither scheme as a function of the frequency ω.
(ii)   Plot it, for values of ω between 0 and $1/4$.
(iii) Write a routine which determines the artifact amplitude as a function of ω for a given binary or ternary scheme.
(iv)   Write a routine which calls the previous one to plot the artifact over the range of ω from 0 to $1/4$.

21.8 Summary

(i)   Using test data sampled from a sinusoid gives some idea of how large the components of the limit curve are which cannot be justified by the original data.

(ii) Calculation of these magnitudes is relatively simple, in terms of the mask and the unit row eigenvector(s) of the scheme.

22. Normalisation of Schemes

Promenade

It was mentioned in earlier chapters that some of their results are inaccurate, that results are shown by methods which apply only to 'normalised' schemes. Here we look at unnormalised schemes and how their properties can be deduced by normalising them. Because there are no obvious advantages in unnormalised schemes, they will be ignored in subsequent parts of this book. The reader impatient to get to the Design chapters can therefore skip this one.

22.1 Quasi-B-Splines

A slightly surprising result is that the limit curve of polygon $P(z)$ under the binary scheme with mask $2\sigma^k(1 + z^2)/2z$ is identical to the limit curve of polygon $\sigma P(z)$ under the scheme whose mask is $2\sigma^{k+1}$. It is a B-spline curve, but with a different control polygon.

This can be traced back to the result that the two-slanted sampling matrix S encountered in the chapter on artifacts has the same product when multiplied on the right by the circulant matrix corresponding to $(1 + z)$ as when multiplied on the left by that corresponding to $(1 + z^2)$.

M. Sabin, *Analysis and Design of Univariate Subdivision Schemes*, Geometry and Computing 6,   133
DOI 10.1007/978-3-642-13648-1_22, © Springer-Verlag Berlin Heidelberg 2010

$$
\frac{1}{2}
\begin{bmatrix}
\ddots & & & \\
1 & 0 & 1 & \\
 & 1 & 0 & 1 \\
 & & 1 & 0 & 1 \\
 & & & & \ddots
\end{bmatrix}
\begin{bmatrix}
\ddots & & & \\
2 & & & \\
0 & 0 & & \\
0 & 2 & & \\
 & 0 & 0 & \\
 & 0 & 2 & \\
 & & & \ddots
\end{bmatrix}
$$

$$
=
\begin{bmatrix}
\ddots & & & \\
2 & & & \\
0 & 0 & & \\
0 & 2 & & \\
 & 0 & 0 & \\
 & 0 & 2 & \\
 & & & \ddots
\end{bmatrix}
\begin{bmatrix}
\ddots & & & \\
1 & 1 & & \\
 & 1 & 1 & \\
 & & 1 & 1 \\
 & & & 1 & 1 \\
 & & & & \ddots
\end{bmatrix}
/2
$$

Remember that circulant matrices commute. We can therefore think of the $1+z^2$ matrix as being the first to be applied to the result of sampling, and this produces the same result as applying $1+z$ to the original polygon first and then sampling. The same move can happen at the end of each refinement step, with the $1+z^2$ of the second step becoming a final $1+z$ in the first. Because there is no end to the refinement process, all that we see is the $1+z$ appearing at the start of it and the changing of an initial $1+z^2$ operation within each step being replaced by a final $1+z$.

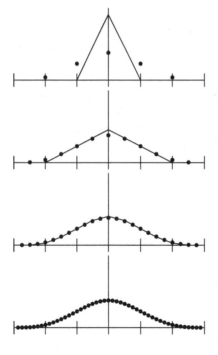

For example, we show here the first three refinements of cardinal data using the mask whose generating function is $2((1+z)/2)^2((1+z^2)/2)^2$, and the argument above says that this has the same limit curve as applying the mask $2((1+z)/2)^4$ to the polygon with vertices $[1,2,1]/4$.

The polygons are shown as lines: the dots are at points of the limit curve, using a unit row eigenvector obtained by convolving that of the normalised scheme $[1,4,1]/6$ with the **prefix** $[1,2,1]/4$.

The limit curve does indeed consist of cubic spans, meeting C^2.

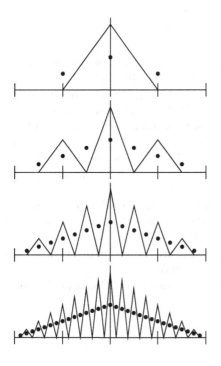

If, however, we look at the second divided difference scheme $2((1+z^2)/2)^2$, we find a more complicated story.

Again, the polygons are shown as lines and the limit points, using an appropriate unit row eigenvector, as dots. We see that the limit points do what we expect, but the polygon shows no sign of converging towards them.

It does not actually diverge, in the way that a scheme with a high norm might be expected to, but its behaviour is uncomfortable.

22.2 Similar effects

Similarly, a scheme containing a factor of $(1 + z^4)$ has the same structure of limit curve as one containing a factor of $(1 + z^2)$ instead, with the original polygon being modified by a $(1 + z^2)$ factor before the subdivision is started, and therefore the same as one containing a factor of $(1 + z)$, with the original polygon being modified by a factor of $(1 + z)(1 + z^2) = (1 + z + z^2 + z^3)$ before subdivision.

This effect is not, of course, limited to schemes generating B-splines.

$$[-1, 1, 7, 9, 9, 7, 1, -1]/16$$

is a scheme which interpolates the midpoints of the edges of the initial polygon. It is not as simple as $[1, 3, 3, 1]/4$, but it is also less expected. Its limit curve is exactly the same as the limit of the four-point scheme applied to the polygon of those midpoints.

Nor is it limited to factors of the form $(1 + z^{2^k})$. Any factor which is a function of only even powers of z has a similar property.

22.3 Summary

(i)   If the generating function of a binary scheme has any factors which are polynomials in z^2, the limit curve is that of its **normalised** form, in which each such factor $q(z^2)$ is replaced by $q(z)$, and a **prefix** (also of $q(z)$ is applied to the initial polygon before refinement starts.

(ii)   This means that the sections above covering generation degree and piecewise polynomial structure above are inaccurate. They apply only in situations where factors polynomial in z^2 do not occur in the mask, or have been converted into ones without, with an operation on the polygon before subdivision starts.

(iii)   This is not a practical concern when choosing or designing a scheme, because a $1 + z^2$ factor, for example, increases the mask width and the support unnecessarily compared with the equivalent $1 + z$ factor, and avoiding the use of such factors also avoids this issue. However, the analysis story would be incomplete without it.

(iv)   The theory describing this is already present in [CDM91]pp161-165.

23. Summary of Analysis Results

23.1 Exercises

(i) Take all of the routines you have written for the various chapters above, and call them from a routine which, when given a scheme reports:
- the support and the practical supports,
- upper and lower bounds on the Hölder continuity,
- the generation, interpolation and reproduction degrees,
- a plot of the basis function
- a plot of the artifact magnitude as a function of spatial frequency.

23.2 Summary

(i) A range of analyses have been described, all relevant to the performance of a subdivision scheme in one application or another.

(ii) There are certain key ideas which appear more than once.
- The use of standard test data sets: cardinal for support, polynomial for precision set, trigonometric for artifact analysis. It is possible that additional sharply chosen data sets could clarify further important analyses.
- The $\sigma = (1 - z^a)/(a(1 - z))$ factor which allows the difference schemes to be determined, and which turns out to be a key concept in artifact analysis also.
- Row eigenvectors giving points and derivatives on the limit curve.
- Higher powers of a scheme giving sharper bounds on continuity and enclosure.

M. Sabin, *Analysis and Design of Univariate Subdivision Schemes*, Geometry and Computing 6,   137
DOI 10.1007/978-3-642-13648-1_23, © Springer-Verlag Berlin Heidelberg 2010

Part IV. Design

In the previous chapters we have seen how, given a mask, we can analyse various aspects of the behaviour of a subdivision scheme. We now address the much more interesting topic of how to invent a subdivision scheme which has some combination of desirable properties.

We do this by looking at the space of all possible schemes, to see how the properties analysed above are distributed within it, and how, given a desirable property, what the subspace is containing all schemes with that property.

Then all we have to do is to take the intersection of the required subspaces.

This sounds easy, but, as usual in a design context, there are trade-offs to be made. Often the requirements are conflicting. However, the following approach allows it to be visible when this is the case.

- What is the space of all possible schemes ?
- What are the subspaces defined by each of the properties that we know how to analyse ?
- How can these be intersected to give the scheme that satisfies the requirements ?

Finally we look at what can be achieved by changing the rules fundamentally, by considering **non-stationary** subdivision, in which the mask changes from step to step. In particular we consider **geometry-sensitive schemes** where the mask is itself determined locally and at every step from the geometry of the polygon.

24. The Design Space

24.1 Binary Schemes

In the analysis of artifacts above, we observed that the mask of every binary, uniform, stationary scheme can be expressed as the product of a number of $\sigma = (1+z)/2\sqrt{z}$ factors and a further factor called the kernel. Then we saw that the kernel itself can be expressed as the sum of a number of terms, each of which is just a constant times an even power of σ.

$$s(z) = 2\sigma^p \sum_{i=0}^{q} c_i \sigma^{2i}, \qquad \text{where } \sigma = \frac{1+z}{2\sqrt{z}}$$

Note that $\sum c_i = 1$ is necessary in order for the overall scheme to have weighted mean stencils.

If we take the σ^p inside the sum, we can express this as

$$s(z) = \sum_{i=0}^{q} 2c_i \sigma^{2i+p}$$

which means simply that every mask is a weighted mean of B-spline masks, either all primal or all dual.

The space of binary uniform stationary schemes is therefore understandable in terms of two components: one contains all primal schemes, with an odd number of entries in the mask, which is a linear combination of odd degree B-splines: the other all dual schemes, with an even number of entries in the mask, which is a linear combination of even degree B-splines. In principle each is only a countably infinite-dimensional space, and in practice each is only a finite dimensional space because we shall not wish to include B-splines above some maximum degree in order to keep the support limited.

Each component may therefore be viewed as having barycentric coordinates, and any choice of a specific set of barycentric coordinates is the design of a scheme.

The coordinate values themselves become the coefficients c_i giving the terms in the kernel, and so translating from them to the mask itself is fairly trivial arithmetic.

M. Sabin, *Analysis and Design of Univariate Subdivision Schemes*, Geometry and Computing 6, 141
DOI 10.1007/978-3-642-13648-1_24, © Springer-Verlag Berlin Heidelberg 2010

24.2 Higher Arities

In the case of ternary schemes a very similar structure holds. If a scheme is both primal and dual, the structure is exactly the same: a linear combination of B-spline masks. But it can be a linear combination of even and odd degree B-splines, because both are both primal and dual. The number of entries in the mask is always odd.

If it is neither primal nor dual, then by symmetry the kernel contains a factor of $[1, 1]/2$, but the space is still that of linear combinations of schemes which have this factor as well as a varying number of $(1 - z^a)/(1 - z)$ factors.

Quaternary and higher arities are significantly more complex. Quaternary schemes can, of course be created by squaring a binary scheme, but there are others which are not so created. Note that the square of a weighted mean of two binary schemes is not the same as the weighted mean of the squares of those schemes.

24.3 Exercises

(i)   What is the mask of the binary scheme $2(3 - 2\sigma^2)\sigma^4$?

(ii) What is the mask of the ternary scheme $3\sigma^3(4 - 3\sigma)$ and what property can you easily identify ?

24.4 Summary

We can look on any binary scheme as an affine combination of B-splines, and choosing the barycentric coordinates in such a combination is a very convenient viewpoint for designing a scheme to have specific properties.

25. Linear Subspaces of the Design Space

Promenade

Each of our analysis criteria leads to a function of position inside the space of all schemes, and those schemes which satisfy some requirement lie inside a subspace. To achieve some combination of requirements means choosing a point of the design space lying in the intersection of the subspaces. If the desired subspaces do not intersect, then the requirements are incompatible.

In this chapter we identify the properties whose specification limits directly the dimension of the design space to be considered.

These are the support, the generation degree and the interpolating degree.

25.1 Support

It is useful to consider this first, because if there is an upper bound on the support as a design criterion, it can be used immediately to limit the design space to relatively few dimensions.

The support is directly related to the width of the mask, which is that of the highest order B-spline with a non-zero coefficient in the barycentric combination. An upper bound on support (which is the most likely kind of design requirement) therefore immediately tells us how many B-splines we need to consider, and how many coefficients we are able to choose.

If the maximum support width acceptable is w, then the expression for the generic scheme satisfying this constraint becomes

$$s(z) = 2 \sum_{i=0}^{w} c_i \sigma^i$$

where the subscripts, i, are all even or all odd.

M. Sabin, *Analysis and Design of Univariate Subdivision Schemes*, Geometry and Computing 6,   143
DOI 10.1007/978-3-642-13648-1_25, © Springer-Verlag Berlin Heidelberg 2010

25.2 Generation Degree

This is useful to take second, because it limits the design space from the other end. If we require the scheme to be able to generate polynomials of degree d_g then the generic scheme becomes

$$s(z) = 2 \sum_{i=d_g+1}^{w} c_i \sigma^i$$

where, again, the subscripts, i, are all even or all odd.

25.3 Interpolation Degree

25.3.1 Primal binary schemes

We consider binary primal schemes, for which the generic scheme can be represented as

$$s(z) = 2\Sigma_{i=d_g+1}^{w} c_i \sigma^i$$

where the sum is taken over even values of $i = 2j$, say, so that

$$s(z) = 2\Sigma_{j=(d_g+1)/2}^{w/2} c_{2j}(\sigma^2)^j.$$

Now σ^2 can be rewritten in terms of δ^2 and so for even i

$$(\sigma^2)^j = (1 + \delta^2/4)^j$$

which can be rewritten as
$$\left(1 + j\left(\tfrac{\delta^2}{4}\right) + \left(\tfrac{j!}{2!(j-2)!}\right)\left(\tfrac{\delta^2}{4}\right)^2 \right.$$
$$\left. + \left(\tfrac{j!}{3!(j-3)!}\right)\left(\tfrac{\delta^2}{4}\right)^3 \cdots \right)$$

We can therefore write

$$s(z) = 2\Sigma_{j=(d_g+1)/2}^{w/2} c_{2j} \sigma^{2j}(z)$$

$$= 2\Sigma_{j=(d_g+1)/2}^{w/2} c_{2j}\left(1 + \delta^2/4\right)^j$$

$$= 2\Sigma_{j=(d_g+1)/2}^{w/2} c_{2j} \left(\begin{array}{c} 1 + j\left(\tfrac{\delta^2}{4}\right) \\ + \left(\tfrac{j!}{2!(j-2)!}\right)\left(\tfrac{\delta^2}{4}\right)^2 \\ + \left(\tfrac{j!}{3!(j-3)!}\right)\left(\tfrac{\delta^2}{4}\right)^3 + \cdots \end{array} \right)$$

$$\cdot \quad = 2\Sigma_{j=(d_g+1)/2}^{w/2} c_{2j}$$

$$+ \Sigma_{i=(d_g+1)/2}^{w} c_{2j} j \frac{\delta^2}{4}$$

$$+ \Sigma_{i=(d_g+1)/2}^{w} c_{2j} \left(\frac{j!}{2!(j-2)!}\right)\left(\frac{\delta^2}{4}\right)^2 + \cdots$$

The interpolation degree is given by one less than the first (non-zero) power of δ with a non-zero coefficient in this expansion of the symbol, and so for linear interpolation we have effectively no constraint.

For cubic precision we have

$$\Sigma_{j=(d_g+1)/2}^{w/2} j c_{2j} = 0$$

and for quintic precision also

$$\Sigma_{j=(d_g+1)/2}^{w/2} \left(\frac{j!}{2(j-2)!} \right) c_{2j} = 0$$

etc.

These conditions are independent linear conditions on the coefficients c_i, which can be satisfied just as far as there are sufficient coefficients c_i available in the range $d - 1 \leq i \leq w$.

Each condition imposed reduces the dimension of the space of possible schemes by one.

If the number of conditions is enough to reduce the dimension to zero, then the values of the c_{2j} can be found by solving a small linear system.

If it is larger, there is no solution and some requirement must be relaxed.

If it is smaller, then some particular solutions can be found by taking the first few c_{2j} then the next set of consecutive ones and then the next... These will give schemes satisfying the requirements, and because of the linearity of these conditions, any linear combination of them will also satisfy them.

This linear subspace can usefully be made more explicit by taking as the corners of our barycentric combination not the B-splines, but schemes satisfying the interpolation degree constraints.

The interpolation degree can also be expressed in terms of the presentation of the unit row eigenvector as a polynomial in δ^2. Since the unit row eigenvector is often shorter than the mask, this might be slightly advantageous.

25.3.2 Non-primal schemes

The interpolation degree can also be expressed in terms of the presentation of alternate terms of the product of the unit row eigenvector with the mask as a polynomial in δ^2. For non-primal schemes this approach has to be taken, since for such a scheme interpolation only means that the limit curve interpolates the data. Vertices of refined polygons do not coincide with original vertices.

Because the unit row eigenvector is not a linear function of the mask, the quasi-interpolation degree is not necessarily a linear constraint in design space.

25.3.3 Higher arities

The situation gets considerably more complex for higher arities.

However, for primal schemes of any arity, any scheme with interpolation degree d_i will leave original vertices unchanged if they lie on a polynomial of degree less than or equal to d_i. A weighted mean of equal points is always the same point, and so this condition remains a linear one.

25.4 Exercises

(i)   How many dimensions does the space of schemes with support ≤ 6 and interpolating degree $d_i \geq 3$ have ?

25.5 Summary

(i)   Provided that the requirements stated are not too ambitious, it is possible to find a set of linearly independent schemes with a given support, generation degree and interpolation degree. This set will typically be quite small, and barycentric coordinates giving linear combinations of them will typically span a space of low dimension.

(ii)   Searching for schemes satisfying further requirements is a relatively easy search process when carried out in a low dimension space of schemes guaranteed to satisfy some of our conditions, compared with searching in the full space for all conditions at once.

26. Non-linear Conditions

Promenade

Once the dimension of the space to be considered has been reduced to a small enough number, the ranges of the barycentric coordinates for which other properties take acceptable values can be determined by a simple search. We are aided in this by the fact that most of these properties are continuous with respect to the c_i, and so a relatively coarse sampling is adequate to give a good idea.

If the dimension is 1, a simple scanning and the plotting of graphs gives a complete enough guide, and we do this here for the case of binary schemes combining two successive B-splines of the same parity.

26.1 Derivative Continuity

The Hölder continuity is a function of the scheme and therefore of the co-efficients when a scheme is expressed as a linear combination of B-splines. For schemes with small kernels it is possible to use sharp specific arguments to determine this directly, at least for certain ranges of coefficients. We saw above that the continuity degree was determined by the kernel, almost inde-pendently of the number of σ factors, which merely added a separate term.

Again we look first at the linear combination of just two consecutive B-splines. The scheme is

$$s(z) = \sigma^d k(z)$$

so that the Hölder continuity is given by

$$d - \log_a(J(k))$$

where $J(k)$ is the joint spectral radius as described in chapter 18 at page 109 above.

For binary schemes the kernel can be written as

$$k(z) = 2((1-c) + c\sigma^2)$$

$$= 2(1-c) + 2c\frac{z^{-1} + 2 + z}{4}$$

$$= 2 - 2c + cz^{-1}/2 + c + cz/2$$

$$= (c/2)z^{-1} + (2-c) + (c/2)z$$

The piece of the matrix we have to look at for eigenanalysis is

$$\begin{bmatrix} c/2 & c/2 \\ & (2-c) \\ & c/2 & c/2 \end{bmatrix}$$

This has eigenvalues $2-c$ and $c/2$ by inspection. It also has a norm of the maximum of $|c|$ and $|2-c|$. We can plot these against c,

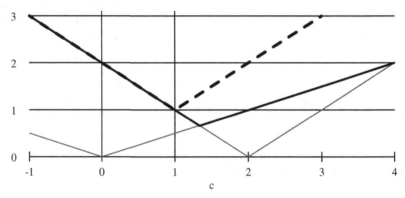

and it becomes apparent that for $c < 1$ the norm and the larger eigenvalue are both equal to $2-c$. Thus $J(k)$ also takes this value.

When c exactly equals 1, the same applies, but in fact the largest eigenvalue becomes that of a polynomial component, and we can take a further two factors of σ out of the kernel. There is an isolated anomalous value here.

For values of c greater than 1, the norm is greater than the largest eigenvalue, and so we need to explore further by taking powers. The square of the scheme is

$$[c^2/4, c(2-c)/2, c(4-c)/4, (2-c)^2, c(4-c)/4, c(2-c)/2, c^2/4]$$

and so the matrix is

$$\begin{bmatrix} c^2/4 & c(4-c)/4 \\ & (2-c)^2 \\ & c(4-c)/4 & c^2/4 \\ & c(2-c)/2 & c(2-c)/2 \\ & c^2/4 & c(4-c)/4 \\ & & (2-c)^2 \\ & & c(4-c)/4 & c^2/4 \end{bmatrix}$$

The row sums of the rows here depend on the value of c.

$1 < c < 2$	$2 < c < 3$	$3 < c < 4$	$4 < c$
c	c	c	$c(c-2)/2$
$(2-c)^2$	$(c-2)^2$	$(c-2)^2$	$(c-2)^2$
c	c	c	$c(c-2)/2$
$c(2-c)$	$c(c-2)$	$c(c-2)$	$c(c-2)$
c	c	c	$c(c-2)/2$
$(2-c)^2$	$(c-2)^2$	$(c-2)^2$	$(c-2)^2$
c	c	c	$c(c-2)/2$

The dominant norm is c in the first two columns and $c(c-2)$ in the third and fourth. To make these values comparable with the figure above we need to take the square roots, giving $\sqrt{c}$ and $\sqrt{c(c-2)}$ respectively.

The top three rows and the bottom three rows are just the square of the original matrix, and so the eigenvalues are just the squares of the originals. We do not need to consider these. By symmetry we can focus on just the 2×2 component

$$\begin{bmatrix} c(4-c)/4 & c^2/4 \\ c(2-c)/2 & c(2-c)/2 \end{bmatrix}$$

whose characteristic equation is

$$4\lambda^2 - \lambda c(8 - 3c) + c^2(2-c)^2$$

The eigenvalues are

$$c\left(8 - 3c \pm \sqrt{c(16 - 7c)}\right)/8$$

Clearly if $c > 16/7 \approx 2.29$ the eigenvalues will be complex and we need to take the modulus which is given by $c(c-2)/4$.

Again we need to take the square root of the modulus of the largest eigenvalue to make it comparable with the original scheme.

Adding these bounds to the previous figure gives

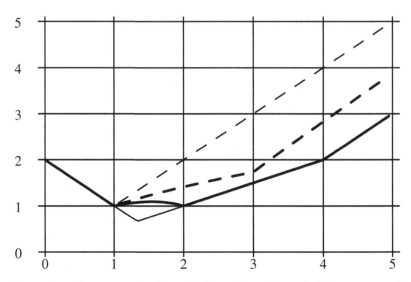

It is clear that going further algebraically will probably give very little insight. Numerically we can take much higher powers, and check both the upper and lower bounds.

Taking the 16th power by repeated squaring gives a scheme of arity 65536. Plotting the upper and lower bounds for each squaring (arities 2, 4, 16, 256 and 65536) makes it look as though the upper bound derived from the norm tends monotonically to the limit determined from the eigenanalysis of relatively low powers.

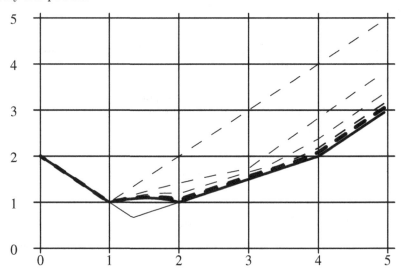

This figure does include five lower bounds, but apart from the interval between $c = 1$ and $c = 2$, where the lower bound comes from the square of the scheme, all are coincident with the bounds from the scheme itself.

It is of interest to note that schemes containing the kernel $[1, 0, 1]$, at $c = 2$, are in fact unnormalised schemes. Their continuity is exactly that determined by the joint spectral radius analysis of the kernel without any need for normalisation.

26.2 Positivity

26.2.1 Another sufficient condition

If all the coefficients c_i are non-negative, then the kernel will have non-negative entries and so the mask will also, and the basis function will be non-negative.

This is a nice trivial result to make very safe design decisions easily. Unfortunately it is so weak that it significantly over-constrains the design space.

26.2.2 Realistic conditions

In the case where just two B-splines are involved, we see directly from the above that for $0 \le c \le 1$ the scheme will be positive.

It is also clear from the first necessary condition that the coefficient of the widest B-spline needs to be positive: in this case $c \ge 0$. Otherwise the extreme entries in the mask would be negative[31].

Above $c = 1$ we need to resort to numerical methods, and we may just as well go direct to what really matters, the l_∞ norm, which dictates by how much the bounding box must be expanded.. Because the l_∞ norm depends on the number of σ factors, we present the result as a set of graphs, covering the cases $2 \le d \le 6$.

---

[31]For situations with more B-splines involved there is a more complicated rule covering the case where the coefficient of the widest B-spline is zero.

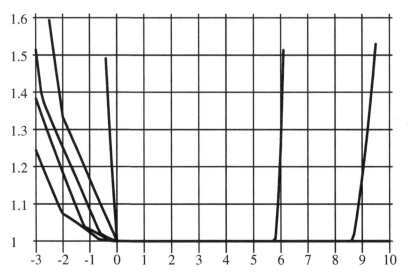

If a certain value of l_∞ is regarded as acceptable, the bounds on c for a given number, d, of σ factors can be read off. The facetting is not an artifact of the plotting: the curves for $c < 0$ really are piecewise linear, and so exact bounds for given values of d and l_∞ are accessible.

26.3 Artifacts

There are two reasonable design criteria based on artifacts. The first is to make the artifact component zero at some specific spatial frequency: the other is to minimise the artifact component over a given range, for example from $\omega = 0$ to $\omega = 1/4$.

26.3.1 Zero artifact at a given spatial frequency

The first is very simple to arrange. We just need to choose the coefficients so that the kernel has a zero at the value of σ which corresponds to the frequency at which we wish to have zero artifact.

Suppose that we want to have zero artifact at $\omega = \omega_0$. Then we require that $k(\sin(\pi\omega_0/2)) = 0$

In the two component case, the kernel is $k(z) = 2((1 - c) + c\sigma^2)$ and this will have a zero when $(1 - c) + c\sin^2(\pi\omega_0/2) = 0$. We determine directly that

$$c = \frac{1}{1 - \sin^2(\pi\omega_0/2)} = 1/\cos^2(\pi\omega_0/2).$$

Note that this value will be greater than 1, tending to 1 as $\omega \to 0$. Also it will never be greater than 2 for $\omega_0 < 1/2$.

$1/n$	ω_0	c
$1/3$	0.333	1.333
$1/4$	0.250	1.172
$1/6$	0.167	1.072
$1/8$	0.125	1.040
$1/10$	0.100	1.025

26.4 Exercises

(i)   The four-point scheme has cubic precision. Identify the scheme with cubic quasi-interpolation degree ($d_i = 3$) with σ^6 as a factor.

(ii)   Determine kernels which give zero artifacts when the number of control points around a circle is 6, 12, 24. How does the coefficient for a given number relate to that at half that number ?

(iii)   Write code to plot the non-polynomial eigenvalue and the norm against the weighted mean coefficient between two arbitrary schemes. Use it to make plots between the two schemes of question (i) above, which have $d_i = 3$.

26.5 Summary

(i)   The non-linear criteria considered above can be evaluated and plotted as functions of the barycentric coordinates in our linear space of schemes.

(ii)   Values of c_i, and thence the entries in an ideal mask, can be read off from these plots.

(iii)   Thus subdivision schemes can be designed to meet required criteria, not just invented and tested.

27. Non-Stationary Schemes

Promenade

All of the above has dealt with stationary schemes where the coefficients in the affine combinations giving refined points remain the same at all steps of the refinement. However, some of the properties we have considered depend primarily on the early steps (such as the support and the artifacts), some on the later ones (such as derivative continuity), and so it seems sensible to consider varying the coefficients between the steps.

We look at two examples which have been suggested in the literature, introduce a new criterion, that of **step-independence**, and open a Pandora's box of schemes which satisfy that criterion.

27.1 Examples

27.1.1 The UP-function.

It appears from our consideration of the support and continuity properties, that there is a serious tension between them. A narrow support, desirable so that separate parts of a curve can be designed fairly independently, requires that the mask have few entries. A high level of derivative continuity demands many $(1 + z)$ factors and thence that the mask be large.

However, the support is dominated by the effect of the first few refinement steps, the continuity by the 'last' few, and so there is the possibility of using a small mask for the first few steps and a larger one later. This is elegantly exemplified by the so-called 'UP' function[32], which is the basis function of a scheme in which at the first step the coefficients used are those of the zero degree B-spline, at the second those of the linear B-spline and, in general, at the nth step those of the degree $n - 1$ B-spline.

This has a finite support, two spans, and infinite differentiability, requirements totally incompatible in the stationary context.

---

[32]The letters 'U' and 'P' are actually Russian characters, whose nearest equivalents are 'I' and 'R' respectively. However the incorrect pronunciation is now used wherever this function is discussed.

The original definition was in terms of a differential equation

$$UP'(x) = 2(UP(2x + 1) - UP(2x - 1))$$

which was invoked by the idea that the derivative of a bell-shaped function should itself look like a bell-shaped function on the left and minus a bell-shaped function on the right. Why not the same bell-shaped functions ?

27.1.2 Variants on UP

In fact UP itself does not work in the parametric curve context, because its support is so narrow. Every span has only two non-zero (positive) basis functions, and so any point lies on a straight line between the two points which influence it. We have just reinvented the polygon, which does not have infinite geometric continuity.

The first variant to look at is therefore to start the process slightly further up the chain of derivatives. If we start at the linear B-spline we get a basis function of support 3, if we start at the quadratic we get one of support 4, and so on. If we call the original UP function UP_0, we can call the others UP_1, UP_2 etc.

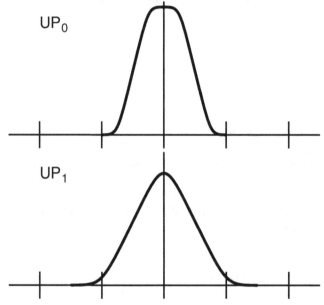

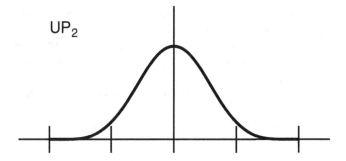

UP_2

27.1.3 STEP UP

Here are two thought experiments.

1)   Instead of carrying out one refinement at each degree in the UP construction, try carrying out two, or three, or several. This fails for UP_0, because the translates of the basis functions no longer sum to 1, but it is OK for UP_1 and UP_2. At each increase in the number of refinements at each degree the support drops, but the continuity remains infinite.

2)   Instead of starting with low degree refinements, carry out the first hundred of whatever scheme you like to name - let's say the four-point scheme. Then after one hundred steps switch to the UP_1 rules. The result will be your scheme for all practical purposes, but the limit curve has infinite continuity. This is an argument which says that Hölder continuity of itself is not an important criterion.

27.1.4 DOWN

Although it seems logical to use a narrow scheme for initial steps and a wider one for later, in fact the exact opposite makes sense in one context, where some specific scheme is used for a number of steps, of whatever width is of interest, but thereafter the degree 1 B-spline is used to generate whatever polygon has been reached at that point. We don't usually look at it in that light, but that is the scheme whose limit curve is the polygon drawn after a number of steps.

27.1.5 Circle-preserving schemes

It was shown, in chapter 21.2 above, that with a stationary scheme non-zero artifacts can be achieved after the first refinement at a finite number of imposed spatial frequencies. However, at the second refinement the original spatial frequency will appear to have halved, relative to the density of the control polygon, and so an infinite number of zeroes in the artifact as a function of spatial frequency would be required to preserve circles exactly.[33]

With a non-stationary scheme, however, for a given initial spatial frequency of polygon, (i.e. a fixed known number of vertices per complete cycle, or vertices forming a regular polygon), the coefficients can vary at each step so that the halving relative frequency of the signal is tracked by a zero of the kernel.

This tracking is regular enough that the updating of the coefficients can be made into a regular recurrence.

27.2 Analyses of Non-Stationary Schemes

27.2.1 Support

UP is definitely an exception. Most of the interesting non-stationary schemes can be looked at as letting the coefficients of a scheme take some trajectory in the design space of fixed finite dimension (and fixed arity) considered above. Except when the coefficient of the widest box-spline happens to drop to zero, the support will remain constant, at that given by the widest scheme included in the linear combinations.

The design of a non-stationary scheme can be regarded then as the design of a **trajectory** in design space.

Such a trajectory has an important property. It could converge towards a limiting scheme, it could repeat itself, or it could wander around randomly. Discount the last possibility. If it repeats itself after a fixed number of steps, then we can take that cycle of steps together as a single scheme of high arity and discover that it is essentially a stationary scheme.

If it does converge to a limit scheme, then that convergence is an important property for those criteria which depend most on the late stages.

27.2.2 Reproduction degree

The simple result here is that each of the polynomial degrees we considered is no lower than the lowest of the individual steps. If a given generation or interpolation degree is required, then the trajectory of the scheme should lie entirely within the set of schemes that has that degree.

In the original work on the four-point scheme, (and much of the later exploration), schemes which are a linear combination of $[0,1,2,1,0]/2$ and $[-1,0,9,16,9,0\ -1]/16$ were considered, the exact linear combination being defined by a **tension parameter**. These schemes were all interpolating.

---

[33]This could in principle be achieved by making the basis function $\pi \sin x/2x$, but this would require a mask of infinite width.

27.2.3 Continuity

It would appear that, if the coefficients are different at every step, there is no hope of applying any of the methods explored earlier for determining the continuity of the limit curve. In fact, however, it makes sense for the trajectory to take a path which converges to a limit, so that after a certain number of steps the scheme becomes almost stationary. When this happens we call the limit the **limit scheme**.

Because the continuity is dominated by the later steps (just think of all the earlier steps as merely making a polygon to which the later steps are applied), if the scheme converges fast enough to its limit scheme, then the continuity properties are indeed those of that limit.

How fast is necessary ? This depends on the degree of continuity being aimed at. Twice as fast as that degree (e.g. quartic convergence to get C2) is provably enough.

27.2.4 Positivity

Positivity analysis is riddled with loose sufficient conditions. Another such is the condition that if every step is positive, however different they may be, the sequence will remain positive. Clearly if no step can create a negative value in the net matrix, it remains positive. We can therefore expect that the overall value of the norm of the scheme will be no worse than the largest value of any of the steps.

27.2.5 Artifacts

The artifact behaviour tends to be dominated by the earlier steps, because after each step the intended spatial frequency is halved relative to the density of vertices along the polygon. Because low spatial frequencies have little artifact effect, the later steps do not cause much distortion.

The early part of the trajectory can be designed to minimise these effects and later parts optimised for some other criterion.

The circle-preserving schemes described above are essentially just carrying out this recipe.

27.3 Step-independence

Non-stationary schemes where the trajectory is pre-defined have a big inelegance: that if one step of refinement is made, perhaps to permit the addition of short-wavelength features, then either the implementation has to remember that the first step has already been made, and start instead at the second, or else, if the implementation starts again with the coefficients of the first stage, those parts of the curve which have not been edited will change their shape. Such changes may be slight and subtle, but they are thoroughly inelegant.

We therefore have an important property, that of **step-independence**, to consider. A scheme is step-independent if the original polygon and the n-times refined polygon have the same limit curve. I.e., if for all values of n they have the same limit curve without the implementation knowing the value of n.

If the progress along the trajectory is not defined by the step-number, what can it be defined by ? The only other option is for it to be defined by the shape of the polygon itself. The scheme becomes **geometry sensitive**. This opens a complete new can of worms, in that the different parts of a given polygon may be suggesting different amounts of progress along the trajectory. This means that such schemes may be not just non-stationary, but also non-uniform. They are worth a chapter in their own right.

27.4 Exercises

(i)   Confirm the support widths of UP_0, UP_1 and UP_2 from the subdivision definition.

(ii) Determine how the coefficients of a dual scheme, a weighted mean of 2σ and $2\sigma^3$, should vary so that a square initial polygon should have a circular limit curve.

27.5 Summary

(i)   A scheme is non-stationary if the coefficients used in the refinement vary from step to step. This is a powerful way of combining good values of properties which depend mainly on the early steps with otherwise incompatible good values of other properties which depend mainly on the later steps.

(ii) A nice way of looking at non-stationary schemes is that they are defined by a trajectory in design space. Both the trajectory and the rate of progress along it can be designed deliberately to give a desired overall behaviour.

(iii) A particularly important aspect of the trajectory is the scheme towards which it converges as the number of steps taken increases without bound.

28. Geometry Sensitive Schemes

Promenade

The step-independence criterion sounds as if it demands that the scheme be stationary, but in fact it does not. It demands that the coefficients should not depend on the step-number, but they can still depend on the shape of the polygon.

The two important avenues of exploration so far have been based on using either the distances between consecutive vertices (**span-based** criteria) or the angles at vertices (**vertex-based** criteria).

28.1 Span Criteria

When the interpolating cubic spline was the standard curve form in CAD systems, it was found that the uniform spline behaved very badly when the points being interpolated were unevenly spaced. There was a perfectly good theory of non-uniform interpolating splines, and various methods were tried of making the knot intervals depend on the distances between points. Actually making them equal to those distances (**chord-length** knot spacing) was not ideal either, and a happy medium, of using the square roots of the distances as knot intervals was found to work quite well. This was called **centripetal** knot spacing.

Uniform B-splines are much less sensitive to poor spacing of the control points, because the first derivative is given by a lower degree B-spline with control 'points' the first differences of the control points. If the control points have first differences all more or less in the same direction, the curve cannot kink back on itself.

The first derivatives can, of course, vary fairly wildly in magnitude if the first differences do, and there is a perfectly good theory of knot-insertion into non-uniform B-splines and this can indeed be expressed in terms of choosing the coefficients in a subdivision implementation.

However, the step-independence criterion is still a relevant one, and, if we insist on it, the knot intervals have to be determined from local data at every

M. Sabin, *Analysis and Design of Univariate Subdivision Schemes*, Geometry and Computing 6,
DOI 10.1007/978-3-642-13648-1_28, © Springer-Verlag Berlin Heidelberg 2010

refinement. This means that our limit curves are not exactly the non-uniform B-splines defined by the control point distances in the original polygon.

There is a lot of subtlety here. Although it is still possible to define a basis function as being the effect of moving one point by an infinitesimal amount, this applies only to one specific original polygon. As soon as you start editing it, the basis functions change.

This means that proofs really have to be hand-crafted from very fundamental principles. You cannot just wave your arms and say 'linearity' because these schemes are no longer linear, and all the short-cuts which linearity brought can no longer be relied upon.

All this difficulty means that schemes have to be designed to be analysable, and this is an art still being developed.

28.2 Vertex Criteria

Another approach resorts to pure geometrical constructions, rather than weighted means with argued coefficients, to determine the new polygon. This is exemplified by a circle-preserving interpolatory scheme which takes the four-point scheme and observes that the new e-vertices are placed where a quadratic through three points has the average of the second derivatives of quadratics through left and right groups of three points.

A geometric analogue of this puts the new e-vertex on the circle whose curvature is the mean of circles through left and right groups of three points. (The curvature can be defined as a vector and so this is well-defined for curves in three-dimensional space.) If four points happened to lie on a circle initially, the mean would give the same circle and so the new point would lie on it too. If all the initial points lay on the same circle, then the limit curve could never leave that circle.

The issue then was, 'which new point on the circle should be chosen ?'. Here the requirement to be able to analyse gave an answer, that if new points were chosen so that in any few-point locality the spacing converged to uniform fast enough, the limit scheme would be the four-point scheme itself, and continuity properties could be argued in that way. Of course in parts of the curve defined by parts of the initial polygon with sparse points, the final density would be lower than in places defined by dense initial polygon points, but these regions would become separated by larger and larger numbers of intermediate points as refinement proceeded. In fact, making arguments about geometric progression proved adequate to support the proofs and also a well-behaved scheme.

The nice property about schemes defined by geometric constructions in this kind of way is that they do appear to have a kind of stationarity. Not stationarity of coefficients, but stationarity of algorithm, which bodes well for simplicity of the implementation and ease of testing.

28.3 Geometric Duality

In classical projective geometry there is a concept of duality slightly different from the one used so far in this book. It is that every theorem or construction which relates points to lines has a dual which relates lines to points. For example, every pair of distinct points determines a line: every pair of distinct lines determines a point.

All of this can be applied easily, at least in the two-dimensional case, to give a complete parallel theory in which a polygon is a sequence of edges (lines) rather than a sequence of points. A new polygon can be created from an old one by taking linear combinations of the lines to make new lines, and the vertices of the new ones just pop out as the places where consecutive new lines intersect.

Such schemes will typically look non-uniform and non-stationary when re-expressed in terms of points, but all the theory is still in fact regular when standing in the dual world.

A nice example is the dual of the quadratic B-spline, which turns into a primal interpolating scheme (because each old line in the old polygon is retained in the new one). The limit curve (the envelope of the lines in the limit polygon) turns out to be a concatenation of conic section pieces.

The problem with this particular duality is that the dual of an inflexion in a curve is an asymptote, so that if the polygon has an inflexion the curve will contain pieces of hyperbola going off to infinity. If rules are put in place that something special is done when the polygon has an inflexion, however, this can give a nice **shape-preserving** scheme which has no inflexions in the limit curve when the initial polygon has none.

In three dimensions the dual of a point is a plane, so that the dual of a curve is a developable surface, the envelope of a univariate set of planes[34].

The dual of a sequence of lines is still a sequence of lines, and so refinement algorithms based on lines[35] can still be set up.

28.4 Summary

(i) Making the coefficients of the subdivision scheme depend on the local geometry allows us to have non-stationarity and step independence at the same time.

(ii) However, it implies non-uniformity, and thus losing the analysis methods which depend on uniformity.

---

[34]All of the subdivision theory remains the same for this interpretation. The only difference is that the numbers on which it operates are interpreted differently in terms of the semantics of the objects they represent.

[35]Possibly represented by Plücker coordinates

(iii) All is not quite lost, because it is possible to argue by proximity when a scheme converges fast enough towards a uniform one as its limit scheme.

(iv) It is also possible to argue that some properties which are maintained at every step will be maintained in the limit also.

Part V. Implementation

Most of this book has been about first analysing a given scheme and then designing a scheme to have some specific attainable compromise between the required properties.

Once a scheme has been designed, however, we need to be able to make use of it. This part therefore addresses the way in which code can be written which allows the resulting curves to be used for practical purposes.

We consider first how to implement the refinement process itself, then how to draw the curve defined for a given scheme and a given initial polygon, and then how to compute the primitive operations used within, for example, a Computer Aided Design software system.

Finally the issues are considered of what end conditions to support, and whether to offer the application of preliminary modifications to the control polygon to make the overall system more ergonomic for the curve designer.

29. Making Polygons

The obvious thing to want to do, given a scheme, is to be able to draw its polygons at successive levels of refinement. Although this is an obvious requirement, it is far from the whole story, and subsequent chapters in this part will address the real requirements. However, as a preliminary to that, we shall consider some different ways of programming the refinement steps themselves, making polygons, given the scheme and the initial polygon.

The first three of these are each obvious, but they differ, depending on the way that you really look at the refinement equation.

29.1 Pull

The first is obvious if you think primarily of the stencils.

```
for each new vertex
do    clear it to zero
      for each entry in its stencil
      do    multiply an old point by that entry and add it in
      od
od
```

We are making each new vertex by pulling the contributions from the appropriate old vertices.

29.2 Push

The second is obvious if you think primarily of the mask.

```
for each new vertex
do    clear it to zero
od
for each old vertex
do    for each entry in the mask
      do    multiply it by the old vertex and
```

```
          add it into the relevant new vertex
      od
od
```

We are taking each old vertex and pushing its contributions to a range of new ones

29.3 Multi-Stage

The third is obvious if you think of the mask as being the product of a kernel and smoothing factors.

```
apply the kernel
for each smoothing factor
do    for each local group of vertices
      do    apply the smoothing
      od
od
```

The kernel is applied first as if it were the mask, by either push or pull, and then the resulting polygon is smoothed out iteratively. It is arguable that this is likely to be the least susceptible to inaccuracies from floating point arithmetic.

29.4 Going Direct

If the objective is the polygon after a reasonably high number of refinement steps, it is more efficient to raise the scheme to that power first, particularly if this is done by using squaring where possible. Once a high-arity mask has been built it can be cached, so that subsequent initial polygons can be processed quickly. The ultimate in speed is to implement a change in the initial polygon by just adding the displacement of a vertex on to the relevant part of the refined polygon by just running through the long mask once.

Clearly the 'high' power need only be high enough to give adequate accuracy.

29.5 Going Direct to Limit Points

If what is required is a polygon with points actually on the limit curve, then these can be constructed by multiplying the polygon constructed in the previous paragraphs by the unit row eigenvector.

Again this can be implemented efficiently by applying this process to the high-arity mask just once before it is cached.

29.6 Summary

Polygons resulting from multiple applications of the subdivision process can be built by alternative methods, all fairly simple. Literal application of the subdivision rules may not be the best in all circumstances.

30. Rendering

Promenade

Rendering is the process of creating a graphical image of a curve. There will be complications like drawing the curve in some particular view, or like trimming to fit a viewport, but these are standard graphics operations, and so we can ignore them here. We are really concerned with making a representation which can be fed into the graphics pipeline.

30.1 Polygon Rendering

The simplest way of doing this is, of course, to apply enough refinements and then just send the edges of the polygon to the routines which do the actual drawing. Simplicity of implementation is important, and this approach is recommended for any first implementation of subdivision software.

However, there are more efficient approaches.

30.2 B-spline Rendering

The first of these is to use instead of the polygon after a number of iterations, the B-spline which, as we saw in chapter 19 above, can be used in the limiting process to define the limit curve. With rendering engines which are capable of accepting cubics, such as PostScript, it is sensible to use a cubic B-spline for this purpose.

PostScript[36] requires its cubics to be provided in Bézier form, but the intermediate control points can easily be determined as the 1/3 and 2/3 points of successive edges. The junction points, at the knot values of abscissa

---

[36]PostScript is a 2D system. In the case of stationary, uniform schemes you can project the initial polygon into 2D first before doing any refinement, but because geometry sensitive schemes may not give the same answer after such a projection as they do before, it is probably better not to start down that route. Project just before feeding the points to the graphics.

M. Sabin, *Analysis and Design of Univariate Subdivision Schemes*, Geometry and Computing 6,   171
DOI 10.1007/978-3-642-13648-1_30, © Springer-Verlag Berlin Heidelberg 2010

are just the average of the nearest such edge-points in the before and after edges.

This approach can give a smooth-looking curve with many fewer spans (and so many fewer refinement steps) than the polygon approach.

30.3 Hermite Rendering

However, the B-splines do not converge to the limit curve any faster than the polygon. The order is quadratic in both cases. If actual accuracy of rendering is important, rather than just beauty, there is another way of making a smooth curve out of cubics which is significantly more accurate for a given number of refinement steps.

This is to evaluate the limit curve points corresponding to the control points, using the row eigenvector of unit eigenvalue, and also the first derivatives, using the next row eigenvector. These are then used to make a Hermite cubic interpolant, which converges as the fourth power of the number of refinements.

30.4 What about Non-stationary Schemes ?

A non-stationary scheme does not have the necessary eigenvectors to apply the above directly to the original polygon. However, in cases where the scheme converges adequately fast towards its own limit, the eigenvectors of the limit scheme can be used with good accuracy after a relatively small number of refinements. How many such refinements are needed has to be determined for each scheme individually.

30.5 Summary

(i)   The simplest way of drawing subdivision curves is just to apply a few steps of subdivision to the given polygon.

(ii)   There are ways of making this much more efficient if speed is important.

31. Interrogation

Promenade

If subdivision curves are to be used within, for example, a CAD system, then it is necessary not just to draw them, but to determine points on them fairly accurately for purposes of, for example, finite element meshing.

We treat three examples here:

31.1 Evaluation at Given Abscissa Values

Because of the fractal nature of the definition, it is not possible in general to evaluate the subdivision curve exactly, except at dyadic points with a small denominator. We have to settle for an approximation within some required tolerance. Because that tolerance can be chosen in the light of the precision needs of the application, this is indeed good enough.

Each of the methods described under rendering above can be applied directly to evaluation. The third, Hermite, form is probably most relevant to applications requiring high accuracy. In fact where the second derivative can also be evaluated exactly at dyadic points, a quintic Hermite interpolant can be used to give an even higher rate of approximation.

The number of refinements is first worked out from the required precision and the initial control polygon. These refinements are then carried out, but only in the smallest possible region around the place where the evaluation is to be made. Doing it everywhere requires excessive computation and storage space. The number of control points needed is only the number required for the evaluation of the points and derivatives at the end of the required span.

31.2 Evaluation at the Intersection with a Given Plane

This is a problem much more typical than just evaluating at a known abscissa. There are two approaches that we can use. The first is to scan along the polygon to find the region likely to be relevant. Then that region is refined as

M. Sabin, *Analysis and Design of Univariate Subdivision Schemes*, Geometry and Computing 6, 173
DOI 10.1007/978-3-642-13648-1_31, © Springer-Verlag Berlin Heidelberg 2010

above, and finally the intersection point is determined by iterating evaluation and correcting an estimated abscissa until the required precision is reached.

The other is to use the convex hull property (if necessary the expanded convex hull, or the expanded bounding box) not only to determine the likely region, but to steer the refinement.

31.3 Evaluation of a Point near a Given Point

There are two closely related enquiries. One is to find the point of the curve actually nearest to the given point. The other is to find all of the points where the distance function is stationary, the **foot-points**. The actual nearest point is unique (except in cases of ties), and may be a foot point or else an end of the curve. Several foot-points may exist for a given curve and target point, some of which will be local minima of distance and some local maxima.

The condition that a span of the curve might contain a foot-point is that the set of vectors from the target point to the hull of the span should have an intersection with the set of planes perpendicular to the tangent vectors in the hull of the first derivatives. The tangent hull is defined exactly analogously to the hull of points, but using the first difference scheme instead of the original scheme.

31.4 Summary

Subdivision curves are parametric curves, and they can be incorporated within Computer-Aided-Design systems as such. Although each curve scheme will need its own low-level evaluation process, much of the mechanism required is standard within CAGD theory and within CAD systems.

32. End Conditions

Promenade

The rest of the book has led up to being able to design into a scheme exactly the right behaviour in the interior of a long curve, but for practical purposes what happens at the ends of a finite piece of curve, and how that is controlled is of at least equal importance.

The previous theory is applicable to finite polygons, provided that they are closed, forming loops. This can be useful, but cannot be called a complete theory. Designers need to be able to create curves which start at one chosen place and finish at another. They also need to be able to influence fairly precisely the derivatives of the curve at those places.

32.1 End Conditions

We consider in fact what happens at the start of the curve, because that is marginally easier to illustrate, but the conditions are equally applicable by symmetry to the other end.

The simplest thing which can happen at the end of a curve subjected to a subdivision construction is for the matrix to just stop. It stops by left hand columns corresponding to non-existent old control points being dropped, as are any rows which then do not sum to unity.

Thus

$$\begin{bmatrix} \ddots & & & & \\ & 1 & 6 & 1 & & \\ & & 4 & 4 & & \\ & & 1 & 6 & 1 & \\ & & & 4 & 4 & \\ & & & 1 & 6 & 1 \\ & & & & & \ddots \end{bmatrix} /8 \quad \text{becomes} \quad \begin{bmatrix} 4 & 4 & & & \\ 1 & 6 & 1 & & \\ & 4 & 4 & & \\ & 1 & 6 & 1 & \\ & & & \ddots \end{bmatrix} /8.$$

When this is done at both ends, the matrix becomes a finite one which can be applied to a finite old polygon. The result is a new polygon with

M. Sabin, *Analysis and Design of Univariate Subdivision Schemes*, Geometry and Computing 6, 175
DOI 10.1007/978-3-642-13648-1_32, © Springer-Verlag Berlin Heidelberg 2010

rather fewer than twice as many vertices. The end of the new polygon is not transparently related to the end of the original. It certainly does not have the same parameter value, and so we can say loosely that the new polygon is shorter than the old.

At each subsequent refinement step, a further shortening takes place, and the limit curve is significantly shorter in parameter space than the original polygon.

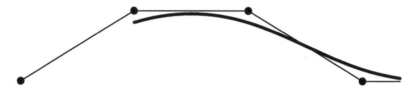

32.2 How Much Shorter?

A useful concept here is that of *the first missing control point*. Once that is articulated, it is clear that the limit curve must lose everything in the support of that first missing control point.

32.3 How do you want the Limit Curve to be Related to the Polygon?

It is far from obvious to the curve designer, who may well want to constrain the position of the end of the curve, how the polygon should be constructed to achieve a particular end-point for the limit curve. It would be much nicer if we could in some way arrange for the limit curve to reach exactly the end of the original polygon.

For the cubic B-spline this can be achieved in an ad-hoc way by just retaining the first control point of the old polygon in the new polygon. This is very easily achieved and implemented, but it has the unfortunate effect that the curvature is always zero at the end of the limit curve.

It would be nice if we could arrange for the derivatives at the end to be reasonably transparently related to the first few control points, and this can be achieved at the price of rather deeper thought about what properties are required and a little more complication in the implementation.

32.4 Requirements for Approximating Schemes

For approximating schemes we can design end-conditions by analogy with the B-splines, where the most widely used end-condition is that called **Bézier end-conditions**.

Strictly speaking these flow simply and naturally from a multiple knot at the end of the domain, and are understood best in terms of unequal interval (non-uniform) B-spline theory, but we can equally well approach them in an ad-hoc fashion, modifying the top left hand corner of the matrix, adding columns as well as rows. For schemes other than box-splines some element of ad-hoc design will be necessary.

The most apparent aspects of these end-conditions are that;-

1   The position of the end of the limit curve is given by the first control point.
2   The first derivative at the end of the curve is given by the first divided difference at the end.
3   The second derivative at the end of the curve is given by the second divided difference at the end.
...   Similarly for higher derivatives up to the generating degree of the scheme.
5   The control points are no longer equally spaced in parameter value.

32.5 Requirements for Interpolating Schemes

The appropriate analogue here is with the various end-conditions which can be devised for interpolating splines[37].

1   The simplest, and probably least useful is the "natural" end-condition, which sets the second derivative to zero at the end of the limit curve.

---

[37]which are not themselves limiting curves of any finite subdivision schemes.

2   Rather better is the **constant curvature end-condition**, which sets the second derivative constant over the first span.

3   For the cubic interpolating spline, there is the **not-a-knot end-condition** which forces the discontinuity of the third derivative at the second knot to be zero, thus making the first two spans part of the same polynomial.

...   For higher degrees there can be imposed zero discontinuities of the relevant derivative at more internal knots, thus forcing the end of the curve to be controlled in a way which is more and more like a Lagrange polynomial.

These lists of properties are a gross oversimplification. Most schemes of interest are neither box-splines nor interpolating schemes, and there is great scope for thoughtful design of the requirements. This has to be a little ad-hoc, because it does depend on the context and on the objectives which the design of the scheme in the interior of the curve tries to meet.

32.6 How to Implement End-conditions

There are essentially three approaches.

- One is to modify the set of stencils at the top of the matrix.

- A second is to modify the original polygon once and for all before starting any subdivision. This usually involves adding extra control points, (whose positions depend on the original ones) but may also involve moving some control points. The new control points can loosely be called **fake points**.

- The third is to modify the polygon before each subdivision step.

The third is obviously equivalent to the first approach because the modification can be expressed as a premultiplication of the old polygon by a matrix, which can alternatively be combined with the standard subdivision matrix to give a modified matrix.

In fact all three approaches are equivalent, and we shall illustrate the first with Bézier end-conditions for box-splines, and the second with Lagrange conditions for the four-point scheme.

Because the second and third approaches rely on the modification of the polygon and the shortening effect during subdivision cancelling each other out, the first approach is numerically preferable. The second involves best separation between the end-condition code and the actual refinement, and so is preferable from the point of view of software robustness.

32.6.1 Modifying the matrices for approximating schemes

Here are the top left hand corners of the matrices for the first few box-spline schemes.

$$\begin{bmatrix} 2 & & & \\ 1 & 1 & & \\ & 2 & & \\ & 1 & 1 & \\ & & & \ddots \end{bmatrix} /2$$

$$\begin{bmatrix} 4 & & & \\ 2 & 2 & & \\ & 3 & 1 & \\ & 1 & 3 & \\ & & & \ddots \end{bmatrix} /4$$

$$\begin{bmatrix} 8 & & & & & \\ 4 & 4 & & & & \\ & 6 & 2 & & & \\ & 1.5 & 5.5 & 1 & & \\ & & 4 & 4 & & \\ & & 1 & 6 & 1 & \\ & & & 4 & 4 & \\ & & & & & \ddots \end{bmatrix} /8$$

The top left corner is always lower triangular. This means that the eigenvalues at the end of the limit curve are given by the diagonal entries, which are successive powers of $1/2$, giving the correct eigenvalue spectrum.

The corresponding row eigenvectors are the same length as the rows, and so the derivatives depend only on the first few old control points at the start of the polygon.

The column eigenvector of the $1/2$ eigenvalue gives the distribution of control points with parameter.

Thus the behaviour at the ends of the curve gets more and more similar to the control at the end of a Bézier curve.

Note that for dual binary schemes the distribution of control points gives an extra half-integer-worth of curve.

32.6.2 Modified initial polygon for interpolating schemes

Here is an example of what can be done by using fake points. Recall that the four-point scheme has a first derivative at a mark point given by the second eigenrow.

$$[1, -8, 0, 8, -1]/12$$

The second derivative is unbounded unless the fourth divided difference happens to be zero, in which case it is given by the second divided difference.

Let the local control points be Y, Z, A, B, C, D

The natural end-condition is given by the simple equations

$$Y := 2A - C$$
$$Z := 2A - B.$$

We can set up end conditions analogous to constant curvature in the end span by asking that the first derivatives at A and B should have the chord $B - A$ as their mean, and that the fourth difference at B should be zero.

$$Y := 34A - 48B + 21C - 6D$$
$$Z := 4A - 6B + 4C - D$$

and conditions analogous to "not-a-knot" by setting the fourth differences at A and B, $Z - 4A + 6B - 4C + D$ and $Y - 4Z + 6A - 4B + C$ to zero. This gives the equations

$$Z := 4A - 6B + 4C - D$$
$$Y := 10A - 20B + 15C - 4D.$$

When this is done the span between A and B is exactly cubic, because of the cubic reproduction property of the 4-point scheme. The first derivative at A is given by

$$18[B - A] - 9[C - A] + 2[D - A]$$

32.7 Summary

(i)   To build software which can provide a convenient medium for the design of curves by placing the vertices of a polygon requires attention to be given to what happens at the ends of the polygon and how conditions at the ends of the limit curve can be controlled.

(ii)   For approximating schemes Bézier end conditions are a useful guide, as are Lagrange or Hermite conditions for interpolating ones.

(iii) However, the detail of designing end conditions does depend on the objectives to which the rest of the scheme is aimed, and so there is great scope for ad-hoc creative thinking.

33. Modifying the Original Polygon

Promenade

The idea of adjusting the original control polygon before starting any refinement was introduced in the previous chapter in the context of end conditions, but it can be used more widely. In particular we can often use an approximating subdivision scheme to interpolate a set of given points.

33.1 Making a Polygon to Interpolate Given Points

The unit row eigenvector is a stencil which gives a point on the limit curve in terms of the original control points. The product of a circulant matrix, E, all of whose rows are equal to that eigenvector, with the control polygon, P, gives a sequence of points, Q, on the limit curve.

$$Q = EP$$

This implies that we could determine a control polygon whose limit curve would interpolate all the points of Q. All we have to do is invert E and multiply Q by it. This is not in fact practical for two reasons. The first is that we do have to worry about end conditions to make E finite. The second is that although E is a narrow-banded matrix, its inverse is typically completely full. It is therefore much cheaper to solve the system $EP = Q$ for P than either to invert E or to multiply Q by it.

The appropriate end-conditions can then be taken in to that solution process.

It is also possible to increase the interpolation degree without going all the way, by using narrow-banded approximations to the inverse of E.

M. Sabin, *Analysis and Design of Univariate Subdivision Schemes*, Geometry and Computing 6,   181
DOI 10.1007/978-3-642-13648-1_33, © Springer-Verlag Berlin Heidelberg 2010

33.2 Summary

(i)   A software system applying subdivision curve theory needs a lot of aspects to be considered which are not usually included in academic papers. Issues like vectorisation and efficient use of multi-core architectures are even beyond the scope of this book.

(ii)   The point being made here is that operations on the polygon as supplied in some way by the curve designer are perfectly legitimate ways for the software creator to make that designer's task easier.

Part VI. Appendices

We include here four appendices. These cover the proofs of four theorems relevant to the body of the book, a short picture of the history of the subject of subdivision curves, worked solutions of the exercises and a coda suggesting interesting topics for future research.

1. Proofs

Theorem 1 The kernel of the square of a scheme is a factor of the square of the kernel

Let $\sigma_a(z)$ denote $\frac{1-z^a}{1-z}\frac{z^{(1-a)/2}}{a}$, so that $\sigma_{a^2}(z) = \frac{1-z^{a^2}}{1-z}\frac{z^{(1-a^2)/2}}{a^2}$ and let the symmetric symbol of a scheme of arity a be $f(z) = (\sigma_a(z))^m k_a(z)$ where $k_a(z)$ is a symmetric Laurent polynomial not divisible by $\sigma_a(z)$, and whose coefficients sum to a.

Then the square of the scheme is given by

$$
\begin{aligned}
f(z)f(z^a) &= (\sigma_a(z))^m k_a(z)(\sigma_a(z^a))^m k_a(z^a) \\
&= (\sigma_a(z)\sigma_a(z^a))^m k_a(z)k_a(z^a) \\
&= \left(\frac{1-z^a}{1-z} \cdot \frac{z^{(1-a)/2}}{a} \cdot \frac{1-(z^a)^a}{1-z} \cdot \frac{(z^a)^{(1-a)/2}}{a}\right)^m \cdot k_a(z)k_a(z^a) \\
&= \left(\frac{1-(z^a)^a}{1-z} \cdot \frac{z^{(1-a)/2}(z^a)^{(1-a)/2}}{a^2}\right)^m \cdot k_a(z)k_a(z^a) \\
&= \left(\frac{1-z^{a^2}}{1-z} \cdot \frac{z^{((1-a)/2+a(1-a)/2)}}{a^2}\right)^m \cdot k_a(z)k_a(z^a) \\
&= \left(\frac{1-z^{a^2}}{1-z} \cdot \frac{z^{(1+a)(1-a)/2}}{a^2}\right)^m \cdot k_a(z)k_a(z^a) \\
&= \left(\frac{1-z^{a^2}}{1-z} \cdot \frac{z^{(1-a^2)/2)}}{a^2}\right)^m \cdot k_a(z)k_a(z^a) \\
&= (\sigma_{a^2}(z))^m \cdot k_a(z)k_a(z^a)
\end{aligned}
$$

This is of the form that we expect if the theorem is true. Indeed, it proves everything that we need for the justification of the algorithms in chapters 17 and 18 above.

However, the question is still open as to whether $k_a(z)k_a(z^a)$ is actually the kernel, because it is conceivable that this could have a further factor of $\sigma_{a^2}(z)$.

M. Sabin, *Analysis and Design of Univariate Subdivision Schemes*, Geometry and Computing 6,   185
DOI 10.1007/978-3-642-13648-1_34, © Springer-Verlag Berlin Heidelberg 2010

Theorem 2 The kernel of the square of a scheme is the square of the kernel

Because $\sigma_{a^2}(z)$ is divisible by $\sigma_a(z)$ and by $\sigma_a(z^a)$ then for there to be such a further factor $k_a(z)k_a(z^a)$ would need also to be divisible by them. We know that by definition $k_a(z)$ is not divisible by $\sigma_a(z)$ and that $k_a(z^a)$ is not divisible by $\sigma_a(z^a)$, and so it would be necessary for $k_a(z)$ to be divisible by $\sigma_a(z^a)$ and $k_a(z^a)$ to be divisible by $\sigma_a(z)$. The first of these is certainly possible, but the second is not, at least for even values of a. This is because if it did have such a factor it would have a root at $z = -1$. Now if $z = -1$ then z^2 and all even powers of z are equal to 1, and the only way for $k_a(z^a)$ to be zero is to have its coefficients sum to zero. They have to sum to a and so for even a, the square of the kernel of the scheme is indeed the kernel of the scheme. This proof needs completing by inclusion of the case for odd a.

Theorem 3 Contribution to the joint spectral radius from a shared eigenvector

Consider the case where A and B are two $n \times n$ matrices which share an eigenvector V.

$$\text{Clearly } AV = \lambda_a V$$
$$\text{and } BV = \lambda_b V$$
$$\text{so } ABV = A\lambda_b V = \lambda_b AV = \lambda_b \lambda_a V$$
$$\text{and } BAV = B\lambda_a V = \lambda_a BV = \lambda_a \lambda_b V$$

Thus V, as well as being an eigenvector of A^2 and B^2, is also an eigenvector of AB and of BA, with eigenvalue $\lambda_a \lambda_b$, which will not be larger than the maximum of λ_a^2 and λ_b^2.

All product sequences of A and B of length l will therefore share the eigenvector V, with an eigenvector which is a weighted geometric mean of λ_a^l and λ_b^l, which will not exceed the larger of λ_a^l and λ_b^l. The joint spectral radius of A and B will not, therefore, come from any mixed sequence of A and B.

Note that this says nothing about contribution from other eigencomponents.

Theorem 4 Contribution to the joint spectral radius from nested invariant subspaces

Consider the case where A and B are two $n \times n$ matrices which share not only an eigenvector V_1 but also a sequence of invariant subspaces of increasing dimension where each subspace contains all its predecessors. Thus the first subspace is spanned by some vector V_1, the second by V_1 and V_2, the third by V_1, V_2 and V_3, and the mth by V_i, $i = 1 \ldots m$.

Apart from V_1, the actual eigenvectors of the matrices are defined only in so far as they span the subspaces. A must have an eigenvector which is a linear combination of V_1 and V_2, and so must B, but they do not have to be the same eigenvector.

Let the largest subspace be of dimension m, which can be less than n, the size of the matrices. Arrange the eigenvectors of A spanning the subspaces in ascending sequence of i, and then append A's other eigenvectors in any sequence. Do the same for B.

Then use Gram-Schmidt orthonormalisation of either result to construct a set of basis vectors.

With respect to this basis, both of the matrices of eigenvectors have the form

$$\begin{bmatrix} U & V \\ 0 & W \end{bmatrix}$$

where U is upper triangular. By appropriate scaling of the eigenvectors these matrices can be given unit diagonals. The inverses of these, the matrices of eigenrows, must also have the same form, and A and B themselves will have the same shape, (though not the unit diagonals), when expressed with respect to that basis because the product of upper triangular matrices is itself upper triangular.

All product sequences of A and B of length l will have the same shape as A and B, showing that the nest of invariant subspaces is common to them all. The eigenvalues corresponding to the eigenvectors in the nest are the diagonal elements of A and B with respect to the chosen basis and so the corresponding eigenvalues are again weighted geometric means of the eigenvalues of A^l and B^l.

Thus the joint spectral radius of A and B will either be the larger of the spectral radii of A and B, or else come from the part of the spectrum which is not associated with the nested invariant subspaces.

2. Historical Notes

The first mathematical study of what we now recognize as subdivision curves was made by Georges de Rham in 1947[deR47]. He had been asked the equation of a shape resulting from a recursive manufacturing process, and produced the answer that it did not have one.

He also proved that the shape was tangent continuous everywhere but did not have a second derivative. Although his arguments were specific to the question at hand, they are recognisably related to the eigenanalysis approach described above.

Subdivision curves were then independently invented in 1974 by George Chaikin, who presented a recursive construction for drawing curves defined by a polygon at an early CAD conference. His paper did not appear in the proceedings, but was submitted to a journal later[Ch74]. In the meantime, however, Richard Riesenfeld[Ri75] and Robin Forrest[Fo74] had independently worked out from the presentation that this construction *did* have an equation. It was the quadratic B-spline. Knot insertion into B-splines was a hot topic at the time, and so it was rapidly realised that B-splines of other degrees also had subdivision constructions. Jeff Lane and Richard Riesenfeld showed[LaRi80] that all of these constructions could be described in terms of a multi-stage process where higher degrees merely needed more stages within each refinement step.

The next big step came a decade later, when Serge Dubuc and Gilles Deslauriers[DeDu85/DeDu87a], stimulated by the ideas of fractals, explored subdivision constructions giving curves which interpolated the vertices of the control polygon. The four-point scheme was just one of a sequence of schemes using $2n$ existing vertices in the stencil for the new edge-vertices, and these stencils were simply derived by the Lagrange interpolation formula for a polynomial of degree $2n - 1$. This gave the generation degree trivially.

The first of these, with $n = 2$, was independently discovered by Nira Dyn, John Gregory and David Levin[DLG87], who also introduced the idea of a tension factor, so that their four-point scheme was a linear combination (in the sense of chapter 24 above) of [-1,0,9,16,9,0,-1]/16 and [1,2,1]/2. This stimulated much exploration into how the value of the tension factor was related to the level of derivative continuity. The most important innovation in this paper was, however, the method for determining the level of continuity using difference schemes. This was first cast without the use of z-transform notation. [DLG91], which does use that notation, is a great deal easier to follow.

The z-transforms were applied without comment as a standard tool in the text-book by Alfred Cavaretta, Wolfgang Dahmen and Charles Micchelli published in 1992[CDM91], a book which is strictly about multivariate schemes, rather than curves. However, many univariate results can be recovered by just omitting one of the many superscripts and subscripts which adorn all the variables. This book is essential reading for mathematicians, though not particularly accessible for computer scientists.

Dyn, Levin and Gregory recast their results in terms of z-transforms in [DLG91]. This paper has been incredibly influential. Referees for journals still see papers submitted which apply their derivative-at-a-time procedure to new schemes with masks apparently plucked from the air.

The joint spectral radius approach appeared in a paper by Ingrid Daubechies and Jeffrey Lagarias[DaLa91] in 1991, and also in one by Hartmut Prautzsch and Charles Micchelli[PM87] in 1987. Efficient computation is still a hot topic. A strong competitor for the ideas described in section 18.3 above is the depth first search method developed by the team of Ulrich Reif.

Many of the other results are harder to pin down. The community was aware of ideas without specific papers being written on them. For example, quasi-interpolation schemes were known long before the 2008 paper of Kai Hormann and Malcolm Sabin[HoSa08] put together enough specifics to be worthy of a journal paper. The idea of support was well understood long before 2004, when Ioannis Ivrissimtzis sorted out systematically the rules for all arities[ISD04].

Convex hull results were already well understood from B-spline theory, but Ron Goldman and Tony deRose[GodR86] showed in 1986 that they could still be applied to schemes which did not have all coefficients in the mask positive.

Recent topics popular for research in the subdivision community have been

- non-stationary schemes, largely stimulated by seeking methods for exponential and trigonometric splines,
- non-linear schemes, stimulated by the desire for methods preserving properties like monotonicity and convexity from the initial polygon to the limit curve (*'shape preservation'*),
- non-uniform schemes, stimulated by the desires for better shaped limit curves when the available data is badly spread, and by the need for compatibility with the non-uniform B-splines, widely used in CAD systems.

Results in these areas are still appearing and it would have been premature to provide any conclusive summaries of those areas.

The biggest omission from this book, and a deliberate one, has been sub-division *surfaces*. This where most of the work has been done, and often the univariate results are just spin-offs. The reader is referred to the book of Joe Warren and Henrik Weimer [WaWe02] for a general overview of subdivision surfaces, and to that of Jörg Peters and Ulrich Reif [PeRe08] for a more detailed examination of the key question of the less than perfect behaviour of subdivision surfaces around extraordinary points.

3. Solutions to Exercises

Worked solutions are provided only for the questions which are not programming exercises.

1 Dramatis Personae

(i) *For each of the schemes in 11.9(i), write out the stencils of the scheme.*
Use the notation which inserts a '' at the place of an edge-vertex.*

$$[1, 4, 6, 4, 1]/8 \text{ has stencils } [1 \quad 6 \quad 1]/8$$
$$\text{and } [4 \quad * \quad 4]/8.$$

$$[1, 3, 3, 1]/4 \text{ has stencils } [1 \quad * \quad 3]/4$$
$$\text{and } [3 \quad * \quad 1]/4.$$

The symmetry of these two stencils means that often only one of the two is explicitly described.

$$[1, 3, 6, 7, 6, 3, 1]/9 \text{ has stencils } [3 \quad * \quad 6]/9,$$
$$[1 \quad 7 \quad 1]/9$$
$$\text{and } [6 \quad * \quad 3]/9.$$

$$[1, 3, 5, 5, 3, 1]/6 \text{ has stencils } [5 \quad * \quad 1]/6,$$
$$[3 \quad * \quad 3]/6$$
$$\text{and } [1 \quad * \quad 5]/6.$$

$$[-1, 0, 9, 16, 9, 0, -1]/16 \text{ has stencils } [-1 \quad 9 * 9 \quad -1]/16$$
$$\text{and } [0 \quad 16 \quad 0]/16.$$

The latter is not usually made explicit, as the scheme is interpolating.

M. Sabin, *Analysis and Design of Univariate Subdivision Schemes*, Geometry and Computing 6,   193
DOI 10.1007/978-3-642-13648-1_36, © Springer-Verlag Berlin Heidelberg 2010

(ii)   *For each of the schemes in 11.9(i), write out the part of the matrix with non-zero principal diagonal. What complications did you find in interpreting this question ?*

$[1, 4, 6, 4, 1]/8$ has matrix
$$\begin{bmatrix} 1 & 6 & 1 & & \\ & 4 & 4 & & \\ & 1 & 6 & 1 & \\ & & 4 & 4 & \\ & & 1 & 6 & 1 \end{bmatrix}/8$$

$[1, 3, 3, 1]/4$ has matrix
$$\begin{bmatrix} 1 & 3 & & \\ & 3 & 1 & \\ & 1 & 3 & \\ & & 3 & 1 \end{bmatrix}/4$$

$[1, 3, 6, 7, 6, 3, 1]/9$ has matrices
$$\begin{bmatrix} 1 & 7 & 1 \\ 6 & 3 & \\ 3 & 6 & \\ 1 & 7 & 1 \end{bmatrix}/9 \text{ and } \begin{bmatrix} 3 & 6 & \\ 1 & 7 & 1 \\ & 6 & 3 \end{bmatrix}/9$$

$[1, 3, 5, 5, 3, 1]/6$ has matrices
$$\begin{bmatrix} 3 & 3 & \\ 1 & 5 & \\ 5 & 1 & \end{bmatrix}/6 \text{ and } \begin{bmatrix} 1 & 5 & \\ 5 & 1 & \\ 3 & 3 & \end{bmatrix}/6$$

$[-1, 0, 9, 16, 9, 0, -1]/16$ has matrix
$$\begin{bmatrix} -1 & 9 & 9 & -1 & & & \\ 0 & 16 & 0 & & & & \\ 1 & 9 & 9 & -1 & & & \\ & 0 & 16 & 0 & & & \\ & 1 & 9 & 9 & -1 & & \\ & & 0 & 16 & 0 & & \\ & & 1 & 9 & 9 & -1 & \end{bmatrix}/16$$

The expected complication is that the ternary schemes have two matrices each.

(iii) *How can the denominator of a scheme be determined from the arity and the sequence of integers in the 11.9(i) examples ? Equally, how can the arity be determined from that sequence and the denominator ?*

Because each stencil has to be a weighted mean, its entries must add up to 1. The number of stencils is equal to the arity, and so the sum of all values in the mask must equal the arity. To get the denominator, therefore divide the total of the mask numerator entries by the arity.

Similarly, if you have the denominator divide that total by it to get the arity.

2 Support

(i) *For each of the schemes of 11.9(i) above, determine the support.*

scheme	support
$[1, 4, 6, 4, 1]/8$	4
$[1, 3, 3, 1]/4$	3
$[1, 3, 6, 7, 6, 3, 1]/9$	3
$[1, 3, 5, 5, 3, 1]/6$	2.5
$[-1, 0, 9, 16, 9, 0, -1]/16$	6

(ii) *What is the square of [1,3,3,1]/4 ?, what is its arity, what are its stencils, and what is its support ?*

The square of $[1,3,3,1]/4$ is $[1,3,6,10,12,12,10,6,3,1]/16$
Its arity is 4, its stencils are

$$
\begin{matrix}
[1 & & 12 & & 3] \\
[& 10 & * & 6 &] \\
[& 6 & * & 10 &] \\
\text{and} \quad [3 & & 12 & & 1]
\end{matrix}
$$

each divided by 16, and its support is the same as that of $[1,3,3,1]$, namely 3.

3 Enclosures

(i) *Which of the five schemes of 11.9(i) above have non-negative basis functions ?*

Schemes 1,2,3 and 4. Only scheme 5 has its basis function somewhere negative.

(ii) *Identify a sequence of control points lying within the band $-1 < y < +1$, for which the limit curve of the four-point scheme goes outside that band.*

The x coordinates are irrelevant. A possible sequence of y coordinates is

$$[\ldots, -0.9, -0.9, +0.9, +0.9, -0.9, -0.9, \ldots].$$

(iii) *Is it necessary for all mask entries to be non-negative for the basis function to be non-negative ?*

No. The scheme of mask $[1, 2, 1, -1, -2, -1, 1, 2, 1]/2$ has a non-negative basis function despite the presence of a negative value in the mask[38].

---

[38] see [CDM91] page 163.

4 Continuity 1 – at Support Ends

(i) *Applying the methods of this chapter to the schemes in 11.9(i) above is totally trivial, but do it anyway.*

The values of f are 1/8, 1/4, 1/9, 1/6 and -1/16 respectively and the values of the Hölder continuity at the ends of the support are

$$-log_2(1/8) = 2+1$$
$$-log_2(1/4) = 1+1$$
$$-log_3(1/9) = 1+1$$
$$-log_3(1/6) \approx 1+0.63$$
$$-log_2(|-1|/16) = 3+1$$

5 Continuity 2 – Eigenanalysis

(i) *Find the Hölder continuity of the quadratic B-spline scheme [1,3,3,1]/4 at the centre of the spans between control points.*

The matrix is

$$\begin{bmatrix} 1 & 3 & & \\ & 3 & 1 & \\ & 1 & 3 & \\ & & 3 & 1 \end{bmatrix}$$

The symmetric and antisymmetric components are

$$\begin{bmatrix} 4 & \\ 3 & 1 \end{bmatrix}/4 \text{ and } \begin{bmatrix} 2 & \\ 3 & 1 \end{bmatrix}/4$$

and the eigenvalues can be read off immediately as 1,1/2,1/4 and 1/4. The column eigenvectors are

$$\begin{bmatrix} 1 \\ 1 \\ 1 \\ 1 \end{bmatrix} \begin{bmatrix} -3 \\ -1 \\ 1 \\ 3 \end{bmatrix} \begin{bmatrix} 1 \\ 0 \\ 0 \\ 1 \end{bmatrix} \begin{bmatrix} -1 \\ 0 \\ 0 \\ 1 \end{bmatrix}$$

and the rows

$$\begin{bmatrix} 0 & 1 & 1 & 0 \end{bmatrix}$$
$$\begin{bmatrix} 0 & -1 & 1 & 0 \end{bmatrix}/2$$
$$\begin{bmatrix} 1 & -1 & -1 & 1 \end{bmatrix}/2$$
$$\begin{bmatrix} -1 & 3 & -3 & 1 \end{bmatrix}/2$$

The fourth of the column eigenvectors is not polynomial, and the associated eigenvalue is 1/4 and so the Hölder continuity is no better than $-log_2(1/4) = 1+1$.

If you have different denominators for the eigenvectors that doesn't matter at this stage, because we have not yet considered what the right values are. However, if you multiply your matrix of columns by the matrix of rows the answer should be a unit matrix.

(ii) *Find the Hölder continuity at both of the mark points of the ternary quadratic scheme [1,3,6,7,6,3,1]/9 .*

The complete matrix is

$$
\begin{bmatrix}
\ddots \\
& 1 & 7 & 1 \\
& & 6 & 3 \\
& & 3 & 6 \\
& & 1 & 7 & 1 \\
& & & 6 & 3 \\
& & & 3 & 6 \\
& & & 1 & 7 & 1 \\
& & & & 6 & 3 \\
& & & & 3 & 6 \\
& & & & 1 & 7 & 1 \\
& & & & & & \ddots
\end{bmatrix} /9
$$

which contains two different submatrices

$$
\begin{bmatrix}
1 & 7 & 1 \\
6 & 3 & \\
3 & 6 & \\
1 & 7 & 1
\end{bmatrix} /9 \text{ and }
\begin{bmatrix}
3 & 6 & \\
1 & 7 & 1 \\
6 & 3 &
\end{bmatrix} /9
$$

The first of these has exactly the same eigenvectors as the binary quadratic scheme of the previous question, but eigenvalues $1,1/3,1/9,1/9$. Again, the fourth of the column eigenvectors is not polynomial, and the associated eigenvalue is $1/9$ and so the Hölder continuity is no better than $-log_3(1/9) = 1 + 1$.

The second, which gives the neighbourhood of a limit point corresponding to a control point has eigenvalues $1,1/3,1/9$, with column eigenvectors

$$
\begin{bmatrix} 1 \\ 1 \\ 1 \end{bmatrix}
\begin{bmatrix} -1 \\ 0 \\ 1 \end{bmatrix} \text{ and }
\begin{bmatrix} 3 \\ -1 \\ 3 \end{bmatrix} /8
$$

and row eigenvectors

$$
\begin{bmatrix} 1 & 6 & 1 \end{bmatrix}/8 \\
\begin{bmatrix} -1 & 0 & 1 \end{bmatrix}/2 \\
\begin{bmatrix} 1 & -2 & 1 \end{bmatrix}/2
$$

The column eigenvectors are all polynomial, and so there is no constraint on the Hölder continuity here.

(iii) *Find the Hölder continuity of the ternary neither scheme [1,3,5,5,3,1]/6 at the 1/4 point.*

The complete matrix is

$$
\begin{bmatrix}
\ddots & & & & & & & & & \\
& 1 & 5 & & & & & & & \\
& 5 & 1 & & & & & & & \\
& 3 & 3 & & & & & & & \\
& 1 & 5 & & & & & & & \\
& & 5 & 1 & & & & & & \\
& & 3 & 3 & & & & & & \\
& & 1 & 5 & & & & & & \\
& & & 5 & 1 & & & & & \\
& & & 3 & 3 & & & & & \\
& & & 1 & 5 & 1 & & & & \\
& & & & & & \ddots & & &
\end{bmatrix} \Big/ 6
$$

This has two matrices, but one is just the mirror image of the other. These twins will share the same eigenvalues, but will have mirrored eigenvectors.

$$
\begin{bmatrix}
1 & 5 \\
& 5 & 1 \\
& 3 & 3
\end{bmatrix} /6
$$

We cannot apply the symmetric/antisymmetric short cut, but can apply the block matrix one.

$$
\begin{bmatrix}
5 & 1 \\
3 & 3
\end{bmatrix} /6
$$

has eigenvalues $1, 1/3$ and so the complete set is $1, 1/3, 1/6$.

The column eigenvectors are

$$
\begin{bmatrix} 1 \\ 1 \\ 1 \end{bmatrix}
\begin{bmatrix} -5 \\ -1 \\ 3 \end{bmatrix} /4 \text{ and }
\begin{bmatrix} 1 \\ 0 \\ 0 \end{bmatrix}
$$

and the row eigenvectors

$$
\begin{array}{ccc}
[0 & 3 & 1]/4 \\
[0 & -1 & 1]/2 \\
[1 & -2 & 1]
\end{array}
$$

The third column eigenvector is non-polynomial, and so the Hölder continuity is no higher than $-log_3(1/6) \approx 1 + 0.65$.

(iv) *Find the Hölder continuity of the four-point scheme [-1,0,9,16,9,0,-1]/16 at the limit points corresponding to the control points.*

This is a much larger problem. The relevant matrix is

$$\begin{bmatrix} -1 & 9 & 9 & -1 & & & \\ & 0 & 16 & 0 & & & \\ & -1 & 9 & 9 & -1 & & \\ & & 0 & 16 & 0 & & \\ & & -1 & 9 & 9 & -1 & \\ & & & 0 & 16 & 0 & \\ & & & -1 & 9 & 9 & -1 \end{bmatrix} /16$$

The symmetric and antisymmetric components are

$$\begin{bmatrix} 16 & & & \\ 9 & 8 & -1 & \\ 0 & 16 & 0 & \\ -1 & 9 & 9 & -1 \end{bmatrix} /16 \text{ and } \begin{bmatrix} 10 & -1 & \\ 16 & 0 & \\ 9 & 9 & -1 \end{bmatrix} /16$$

with eigenvalues 1,1/4,1/4,-1/16 and 1/2,1/8,-1/16 respectively.

There is a Jordan block from the two 1/4 eigenvalues, which have a coupling value of 1. This causes a new component of the second difference (of size proportional to the fourth difference) to be added at each refinement. Thus the second difference grows arithmetically, and the Hölder continuity is no better there than 1+1.

6 Continuity 3 - Difference Schemes

(i)   *How many continuous derivatives does the ternary neither scheme [1,3,5,5,3,1]/6 have ?*

The difference scheme is $[1, 2, 2, 1]/6$, with stencils $[1, 1]/6$, $[2]/6$ and $[2]/6$ and so the norm is $1/3 < 1$. The scheme is convergent.

The divided difference scheme is $[1, 2, 2, 1]/2$ whose difference scheme is $[1,1]/2$. The stencils are $[1]/2$ and $[1]/2$ and so the norm is $1/2 < 1$. The first divided difference scheme is convergent.

The second divided difference scheme is $3[1, 1]/2$ which does not have any factors of σ and so we cannot make any statement about the second derivative continuity. The second derivative is probably not continuous.

7 Continuity 4 - Difference Eigenanalysis

(i)   *What is σ for a ternary scheme ?*

For a ternary scheme $\sigma = (1 - z^3)/3(1 - z) = (1 + z + z^2)3$

(ii) *What is the kernel of the ternary neither scheme [1,3,5,5,3,1]/6, and how many σ factors does it have ?*

$$\frac{[1,3,5,5,3,1]/6}{[1,1,1]/3} = [1,2,2,1]/2$$

$$\frac{[1,2,2,1]/2}{[1,1,1]/3} = 3[1,1]/2$$

Thus the kernel is $3[1,1]/2$ and the number of σ factors is two.

Both the eigenvalue and the norm of the kernel are equal to $3/2$, and so the Hölder continuity is $2 - log_3(3/2) = 1 + log(2)/log(3) = 1 + 0.630929$.

8 Reproduction of Polynomials

(i)   *Check the reproduction degree of the four-point scheme [-1,0,9,16,9,0,-1]/16.*

This scheme can be expressed as $2\sigma^4(3 - 2\sigma^2)$ and the generating degree is therefore 3.

The v-stencil is $[0,16,0]/16$ which can be expressed as $1 + 0\delta^2 + 0\delta^4 + \ldots$ and so the quasi-interpolating degree is not bounded.

The reproduction degree is the lower of these two and is therefore 3.

9 Artifacts

(i)   *Determine the artifact amplitude of the ternary neither scheme as a function of the frequency ω.*

This is a fairly tough question. Because the ternary neither scheme does not have unit eigenrows for the vertices and mid-spans as it stands, we have to take its square, thus creating a 'both' scheme.

$$[1,3,5,8,12,16,20,24,28,30,30,30,28,24,20,16,12,8,5,3,1]/36$$

When this is expressed as a matrix, the square pieces centred on the vertices and mid spans are

$$\begin{bmatrix} 5 & 30 & 1 \\ 3 & 30 & 3 \\ 1 & 30 & 5 \end{bmatrix}/36 \text{ and } \begin{bmatrix} 20 & 16 \\ 16 & 20 \end{bmatrix}/36$$

of which the unit eigenrows are $[1,10,1]/12$ and $[1,1]/2$ respectively. Writing the second of these as $[6,6]/12$ and combining them we get $[1,6,10,6,1]/12$ as an equivalent binary mask. This is $2[1,2,1]*[1,4,1]/24$ which can be written as $\sigma^2(2 + \sigma^2)/3$.

The expression for the artifact as a function of ω is therefore

$$\sin^2(\pi\omega/2)(2 + \sin^2(\pi\omega/2))/3$$

(ii)   *Plot it, for values of w between 0 and 1/4.*

This plot is not very informative. The value at $w = 1/4$ should be about 0.2235 and the variation close to $w = 0$ should be quadratic.

10 The Design Space

(i)   *What is the mask of the binary scheme $2\sigma^4(3 - 2\sigma^2)$?*
       Write this as $\sigma^4(6 - 4\sigma^2)$

$$\sigma^4 = [1, 4, 6, 4, 1]/16$$
$$4\sigma^2 = [1, 2, 1]$$
$$6 - 4\sigma^2 = [-1, 4, -1]$$

	1	4	6	4	1	
	−1	4	−1			

−1	−4	−6	−4	−1		
	4	16	24	16	4	
		−1	−4	−6	−4	−1

| −1 | 0 | 9 | 16 | 9 | 0 | −1 |

The mask is [-1,0,9,16,9,0,-1]/16.

(ii)   *What is the mask of the ternary scheme $3\sigma^3(4 - 3\sigma)$ and what property can you easily identify ?*

$$\sigma = [1, 1, 1]/3$$
$$\sigma^3 = [1, 3, 6, 7, 6, 3, 1]/27$$
$$3\sigma^3 = [1, 3, 6, 7, 6, 3, 1]/9$$
$$3\sigma = [1, 1, 1]$$
$$4 - 3\sigma = [-1, 3, -1]$$

1	3	6	7	6	3	1		
−1	3	−1						

−1	−3	−6	−7	−6	−3	−1		
	3	9	18	21	18	9	3	
		−1	−3	−6	−7	−6	−3	−1

−1	0	2	8	9	8	2	0	−1

The mask is $[-1,0,2,8,9,8,2,0,-1]/9$ and its scheme has the obvious property that because the entries three from the centre are zero, it is an interpolating scheme.

11 Linear Subspaces of the Design Space

(i)   *How many dimensions does the space of schemes with support ≤ 6 and interpolating degree $d_i \geq 3$ have ?*
The space is of dimension zero, i.e. it contains just one point, which is the four-point scheme.

12 Non-linear Conditions

(i)   *The four-point scheme has cubic precision. Identify the scheme with cubic quasi-interpolation degree $(d_i = 3)$ with σ^6 as a factor.*
We can express σ^6 as a polynomial in δ^2

$64\sigma^6$	1	6	15	20	15	6	1
δ^6	1	−6	15	−20	15	−6	1

	12	0	40	0	12
$12\delta^4$	12	−48	72	−48	12

	48	−32	48
$48\delta^2$	48	−96	48

64

Thus $64\sigma^6 = \delta^6 + 12\delta^4 + 48\delta^2 + 64$. We therefore need to multiply it by a polynomial in δ^2 starting with $(64 - 48\delta^2)/64$ in order to get a zero coefficient for δ^2, and the simplest of these is $(64 - 48\delta^2)/64 = (4 - 3\delta^2)/4$.
This can be expressed as $[-3,10,-3]/4$ or as $(4-3\sigma^2)$ and the mask required is given by the product, $2\sigma^6(4 - 3\sigma^2)$

The strategy used here, of subtracting multiples of high powers first of $(1 - \delta^2)$ is probably the most effective for hand-calculation for small masks. The alternative strategy of noting quotients and remainders for successive division by $(1 - \delta^2)$ is probably a more general way to go for an algorithm which might be faced with masks of any size.

(ii) *Determine kernels which give zero artifacts when the number of control points around a circle is 6, 12, 24. How does the coefficient for a given number relate to that at half that number ?*

We need kernels of the form $(1-c)+c\sin^2(\pi\omega/2)$ for $\omega = 1/6, 1/12, 1/24$. The expression for the kernel will be zero when $\sin^2(\pi\omega/2) = (c-1)/c$, or $c = 1/(1 - \sin^2(\pi\omega/2)) = 1/\cos^2(\pi\omega/2) = 2/(1 + \cos(\pi\omega))$.

This can be tabulated

n	$\pi\omega$	$\cos(\pi\omega)$ (in degrees)	c
6	30	0.866	1.07179
12	15	0.966	1.01734
24	7.5	0.9914	1.00429

Using the double angle formula $\cos(2x) = 2\cos^2(x) - 1$ we can determine that if c is the coefficient for ω, the coefficient c' for 2ω is given by

$$c' = 2\sqrt{c}/(1 + \sqrt{c}).$$

13 Non-stationary Schemes

(i) *Confirm the support widths of UP_0, UP_1 and UP_2 from the subdivision definition.*

A B-spline of degree d has $d+2$ entries in the mask and the distance from the centre of symmetry to the extreme new point is $(d + 1)/4$

Thus B-splines whose degrees $d(l)$ depend on the level l of an UP scheme will add contribute $\Sigma_{l=0}^{\infty}(d(l) + 1)/(4 * 2^l)$ to the distance from the centre of symmetry to the edge of the support.

In the case of UP_0 $d(l) = l$ and so we need to evaluate (to get half the support width)

$$S = \Sigma_{l=0}^{\infty}(l + 1)/(4 * 2^l)$$
$$= 1/4 + \Sigma_{l=1}^{\infty}(l + 1)/(4 * 2^l)$$

Letting $l = j + 1$
$$= 1/4 + \Sigma_{j=0}^{\infty}(j + 2)/(8 * 2^j)$$
$$= 1/4 + S/2 + \Sigma_{j=0}^{\infty}1/(8 * 2^j)$$
$$= 1/4 + S/2 + 2/8$$

Hence
$$S/2 = 1/4 + 2/8$$

or
$$S = 1 \qquad \text{giving a support of 2}$$

The values for UP_1 and UP_2 can be derived by noting that (by making the initial degree 0 B-spline step) we can make the support UP_0 by adding half that of UP_1 (because it is being applied at the second step) to the distance ($=1/2$) between the two new control points.

Thus the support of $UP_1 = 2(2 - 1/2) = 3$ and similarly for UP_2.

(ii) *Determine how the coefficients of a dual scheme, a weighted mean of 2σ and $2\sigma^3$, should vary so that a square initial polygon should have a circular limit curve.*

The new scheme will be of the form $2\sigma((1-c)+c\sigma^2)$, which is very similar to the problem solved above.

All we need to add is the initial value of c, which is given by

$$c = 2/(1 + \cos(\pi w)) = 2/(1 + \cos(\pi/4)) = 2/(1 + 1/\sqrt{2}) = 1.172$$

After the each step the new coefficient c' is given by

$$c' = 2\sqrt{c}/(1 + \sqrt{c})$$

.

4. Coda

A coda is supposed to tidy everything up, so that the reader gets a feeling of rounded completeness. This coda will not do that. At the start of the book it was suggested that one of its purposes was to trigger others to prove that the theory wasn't complete after all. That process already started during the writing of the book, as one aspect after another turned up that would have been nice to include, but was not completely enough understood. Here, therefore, are some of the holes in the story which look as if they might lead to productive insights.

1 Fourier Decay Analysis

Much of the earliest work was done using the ideas of the Fourier transform. In particular, there is a measure of smoothness which is not the same as Hölder continuity but eerily related to it, based on the rate at which the Fourier transform of the basis function decays at high frequencies.

Just as the Hölder continuity can be measured exactly at rational points with a finite computation, but only bounds can be found for the curve as a whole, the Fourier decay rate can be measured exactly at rational frequencies with a finite computation, but only bounds found for the spectrum as a whole. The two measures turn out to be identical for the B-splines, but are typically different (by a small amount) for other basis functions. Both are changed by the same amount for every additional σ factor in the mask.

The Fourier domain arguments do appear to apply well to non-stationary schemes, and they deserve more attention now.

2 Fourier Energy

The artifact analysis above is limited, essentially to binary and ternary schemes. Higher arities can bring in artifacts at higher frequencies which might spoil or improve the shapes of the limit curves, and it is not obvious how these can most sensibly be handled.

M. Sabin, *Analysis and Design of Univariate Subdivision Schemes*, Geometry and Computing 6,   205
DOI 10.1007/978-3-642-13648-1_37, © Springer-Verlag Berlin Heidelberg 2010

The total energy in the Fourier transform above the Shannon limit (or its ratio to that below) might provide a way of measuring artifacts which is focussed on the effects of the first few refinements, but extends naturally into later ones, and thus also covers high arities elegantly.

3 Links between the Artifacts and Approximation Order

There is a tantalising link between the plots at the end of the artifacts chapter and the polynomial degrees in the preceding one. The sum of the artifact and signal curves has a Taylor expansion around $\omega = 0$ which almost ties up with the interpolation degree. The artifact curve itself has a Taylor expansion around $\omega = 0$ which almost ties up with the generation degree. This needs looking at.

4 End-conditions for Schemes with Higher Quasi-interpolation Degree

The end-conditions described above cover two distinct cases, those of interpolating schemes, which are likened to Lagrange interpolation, and those of B-splines, likened to the Bézier end-conditions. The schemes which interpolate when the data lies on a cubic or higher polynomial do not really fit either of these cases. They are almost interpolating (when the data is really smooth) but not quite. Somebody needs to play with these schemes to find out how they currently misbehave at the ends and what kinds of control are required to make them do what the curve designer wants.

5 Non-uniform Theory to Encompass Endconditions

Handling non-uniformity is a known research topic at the time of writing. In the case of B-splines, the end conditions can elegantly be treated as a specific case of non-uniformity. Can this be extended neatly to other subdivision schemes ?

Non-uniformity also brings the question of where within each span each new knot should be inserted. Forcing it always to be at the centre is a totally arbitrary choice. A second question is close behind, whether to insert in every span anyway. Quasi-crystal theory, which inserts knots only one at a time, but where they are most needed, does not lead to obvious efficiency, but could provide some theoretical insight.

Bibliography

[deR47]   G.de Rham: Un peu de mathématique à propos d'une courbe plane.
           Elemente der Mathematik 2, pp73–76, 1947
[ShWe72]   C.E.Shannon and W.Weaver: The Mathematical Theory of Commu-
           nication. University of Illinois Press, (paperback edition) ISBN 0-252-
           72548-4
[Ch74]   G.Chaikin: An algorithm for high speed curve generation. Computer
           Graphics and Image Processing 3, pp346–349, 1974
[Fo74]   A.R.Forrest: Notes on Chaikin's algorithm. University of East Anglia
           Computational Geometry Project Memo CGP74/1, 1974
[Ri75]   R.Riesenfeld: On Chaikin's algorithm. Computer Graphics and Image
           Processing 4, pp304–310, 1975
[deB78]   C. de Boor: A Practical Guide to Splines. Springer 1978
[LaRi80]   J.Lane and R.Riesenfeld: A theoretical development for the computer
           generation and display of piecewise polynomial surfaces. IEEE Trans
           Pattern Anal. Machine Intell. 2(1), pp35–46, 1980
[Du82]   S.Dubuc: Une foire de courbes sans tangentes. pp99–123 in Actualités
           mathematique, Actes VIième Congrès mathematiques d'expression
           latine, Gauthiers-Villars, Paris 1982
[DeDu85]   G.Deslauriers and S.Dubuc Interpolation Dyadique. pp44–56 in Sem-
           inar Hausdorf du 21 Mai 1985.
[CoSc85]   E.Cohen and L.L.Schumaker: Rates of Convergence of control poly-
           gons. CAGD 2(1–3), pp229–235, 1985
[Da86]   W.Dahmen: Subdivision Algorithms converge quadratically.
           J.Comput.Appl.Math 16, pp145–158, 1986
[DyLe86]   N.Dyn and D.Levin: Smooth Interpolation by bisection algorithms.
           pp335–337 in Approximation Theory 5 (eds Chui, Schumaker and
           Ward), 1986
[M86]   C.A.Micchelli: Subdivision algorithms for curves and surfaces. SIG-
           GRAPH 1986
[GodR86]   R.N.Goldman and T.D.deRose: Recursive subdivision without the
           convex hull property. CAGD 3(4), 247–265, 1986
[DeDu87a] G.Deslauriers and S.Dubuc: Transformées de Fourier de Courbes Ir-
           regulières. Ann Sc.Math.Quebec 11(1), pp25–44 1987
[DeDu87b] G.Deslauriers and S.Dubuc: Dyadic Interpolation. chapter 4 in an
           english translation (Wiley 1987) of [DeDu85]
[DLG87]   N.Dyn, D.Levin and J.Gregory: A 4-point interpolatory scheme for
           curve design. CAGD 4(4) pp257–268, 1987
[PM87]   H.Prautzsch and C.A.Micchelli: Computing curves invariant under
           halving. CAGD 4(1/2) pp113–140, 1987
[deB87]   C. de Boor: Cutting Corners always works. CAGD 4, pp125–131, 1987

[MP87]   C.A.Micchelli and H.Prautzsch: Refinement and subdivision for spaces
            of integer translates of compactly supported functions. pp192–222 in
            Numerical Analysis (eds Griffith, Watson), Academic Press 1987

[He88]   M.J.Hejna: Curves constructed by geometrically based algorithms.
            dissertation RPI 1988

[MP89]   C.A.Micchelli and H.Prautzsch: Uniform Refinement of Curves. Lin-
            ear Algebra Appl 114/115 pp841-870, 1989

[DeDu89]   Gilles Deslauriers and Serge Dubuc: Symmetric Iterative Interpola-
            tion Processes. Constructive Approximation 5: 49-68, 1989

[CM89]   Alfred S Cavaretta and Charles A Micchelli: Subdivision Algorithms.
            pp115–153 in Mathematical Methods in Computer Aided Geometric
            Design (eds Lyche and Schumaker), Academic Press 1989 ISBN 0-12-
            460515-X

[deB90]   C. de Boor: Local corner cutting and the smoothness of the limiting
            curve. CAGD 7(5), pp389–398, 1990

[DyLe90]   N.Dyn and D.Levin: Interpolating subdivision schemes for the gener-
            ation of curves and surfaces. pp91–106 in Multivariate Interpolation
            and Approximation (eds Hausman and Jetter), Birkhauser Verlag,
            Basel. 1990

[R90]   V.A.Rvachev: Compactly supported solutions of functional differen-
            tial equations and their applications. Russian Math Surveys 45,1
            pp87-120

[DGL91]   Nira Dyn, John Gregory and David Levin: Analysis of Uniform Binary
            subdivision schemes for curve design. Constructive Approximation 7
            pp127–147 1991

[DLYS91]   N.Dyn, D.Levin and I.Yad-Shalom: Regularity conditions for a class
            of geometrically continuous curves and surfaces. pp169–176 in Curves
            and Surfaces (ed Laurent, Le-Méhauté and Schumaker), Academic
            Press 1991 ISBN 0-12-438660-1

[Sa91]   M.A.Sabin: ω-convergence, a criterion for linear approximation.
            pp411–414 in Curves and Surfaces (ed Laurent, Le-Méhauté and
            Schumaker) Academic Press 1991 ISBN 0-12-438660-1

[CDM91]   A.S.Cavaretta, W.Dahmen and C.A.Micchelli: Stationary Subdivi-
            sion. Memoirs of American Mathematical Society 453, 1991

[DaLa91]   I.Daubechies and J.C.Lagarias: Two scale difference equations I. Ex-
            istence and Global regularity of solutions. SIAM J.Math Anal 22(5)
            pp1388-1410, 1991

[DaLa92]   I.Daubechies and J.C.Lagarias: Two scale difference equations II. Lo-
            cal regularity, infinite products of matrices and fractals. SIAM J Math
            Anal 23, pp1031-1079 1992

[E92]   T.Eirola: Sobolev Characterization of solutions of dilation equations.
            SIAM J Math Anal 23(4), pp1015-1030, 1992

[DyLe92]   N.Dyn and D.Levin: Stationary and Non-stationary subdivision
            schemes. pp209–216 in Mathematical methods in Computer Aided
            Geometric Design (eds Lyche and Schumaker), 1992

[R92]   O.Rioul: Simple regularity criteria for subdivision schemes. SIAM
            J.Math Anal 23, p1544–1576, 1992

[BM92]   M.Buhmann and C.A.Micchelli: Using 2-slanted matrices for subdi-
            vision report 1992/NA4 of Cambridge University Numerical Analysis
            Group.

[M92]   J.-L.Merrien: A family of Hermite interpolants by bisection algo-
            rithms. Numerical Algorithms 2, pp187–200, 1992

[deB93]   C. de Boor: B(asic)-spline basics. pp27–49 in Fundamental Develop-
            ments of Computer-Aided Geometric Modelling (ed Piegl), Academic
            Press 1993

[MeUt94]   A. Le Méhauté and F.I.Utreras: Convexity preserving interpolatory
            subdivision. CAGD 11(1),pp17–38, 1994

[DDL95]   G.Derfel and N.Dyn and D.Levin: Generalized Refinement Equa-
            tions and Subdivision Processes. Journal of Approximation Theory
            80, 272–297, 1995

[DL95]   N.Dyn and D.Levin: Analysis of Asymptotically Equivalent Binary
            Subdivision Schemes. Journal of Mathematical Analysis and Appli-
            cations 193, pp594–621, 1995

[DyGL95]   N.Dyn, J.Gregory and D.Levin: Piecewise uniform subdivision
            schemes. pp111–119 in Mathematical Methods for Curves and Sur-
            faces (eds Daehlen, Lyche and Schumaker) Vanderbilt University
            Press 1995 1995 ISBN 0-8265-1268-2

[Wa95]   J.Warren: Binary Subdivision Schemes for Functions over Irregular
            Knot Sequences. pp543–562 in Mathematical Methods for Curves and
            Surfaces (eds Daehlen, Lyche and Schumaker), Vanderbilt University
            Press 1995 ISBN 0-8265-1268-2

[GrQu96]   J.Gregory and R.Qu: Non-uniform corner cutting. CAGD 13(8),
            pp763–772, 1996

[KaWe97]   D.P.Kacsó and H-J.Wenz: On an Almost-Convex-hull property. pp
            217–222 in Curves and Surfaces with applications in CAGD (eds Le
            Méhauté, Rabut and Schumaker), Vanderbilt University Press 1997
            ISBN 0-8265-1293-3

[PPS97]   M.Paluszny, H.Prautzsch and M.Schäfer: A geometric look at corner
            cutting. CAGD 14(5), pp421–448 1997

[Au97]   G.Aumann: Corner cutting curves and a new characterization of
            Bézier and B-spline curves. CAGD 14(5), pp449–474, 1997

[ZaTo97]   C.E.Zair and E.Tosan: Unified IFS-based model to Generate Smooth
            or Fractal Forms. pp345–354 in Surface Fitting and Multiresolution
            Methods (eds Le Méhauté, Rabut and Schumaker) Vanderbilt Uni-
            versity Press 1997 ISBN 0-8265-1294-1

[RS97]   A.Ron and Z.Shen: The sobolev regularity of refinable functions.
            Preprint 1997

[Wa97]   J.Warren: Sparse filter banks for binary subdivision schemes. pp427–
            438 in Mathematics of Surfaces VII (eds Goodman and Martin) 1997

[Ko98]   Leif Kobbelt: Using the discrete fourier transform to analyze the con-
            vergence of subdivision schemes. Applied and Computational Har-
            monic Analysis, Volume 5(1), pp68-91, 1998

[MaPe99]   E.Mainar and J.M.Peña: Corner cutting algorithms associated with
            optimal shape preserving representations. CAGD 16(9), pp883–906,
            1999

[Le99]   D.Levin: Using Laurent polynomial representation for the analysis
            of non-uniform binary subdivision schemes. Adv. Comput. Math 11,
            pp41–54, 1999

[DyLy99]   N.Dyn and T.Lyche: Analysis of Hermite interpolatory subdivision
            schemes. Spline Functions and the Theory of Wavelets (ed Dubuc)
            AMS series CRM Proceedings and Lecture Notes 18, pp105–113, 1999

[LuPe00]   David Lutterkort and Jörg Peters: Linear Envelopes for Uniform B-
            spline curves. pp239–246 in Curve and Surface Design (eds Laurent,
            Sablonnière and Schumaker) Vanderbilt Press 2000, ISBN 0-8265-
            1356-5

[NOW01]   A.Nasri, K. van Overfeld and B.Wyvill: A recursive subdivision algorithm for piecewise circular spline. Computer Graphics Forum 20(1) pp35–45, 2001

[WaWe02]   J.Warren and H.Weimer: Subdivision methods for Geometric Design. Morgan Kaufmann, 2002 ISBN 1-55860-446-4,

[Dyn02]   N.Dyn: Subdivision Schemes in Computer-aided geometric design. pp36–104 in Advances in Numerical Analysis (ed Light), Oxford University Press 2002

[DyLe02]   N.Dyn and D.Levin: Subdivision schemes in geometric modelling. pp 73–144 in Acta Numerica 11, Cambridge University Press, 2002

[JuSc02]   B.Jüttler, U.Schwaneke: Analysis and Design of Hermite subdivision schemes. The Visual Computer 18, pp326–342, 2002

[HIDS02]   M.F.Hassan, I.P.Ivrissimtzis, N.A.Dodgson and M.A.Sabin: An interpolating 4-point C^2 ternary stationary subdivision scheme. CAGD 19(1), pp1–18, 2002

[JSD02]   M.K.Jena, P.Shunmugaraj and P.C.Das: A subdivision algorithm for trigonometric spline curves. CAGD 19(1), pp71–88, 2002

[IDHS02]   I.P.Ivrissimtzis, N.A.Dodgson, M.F.Hassan and M.A.Sabin: The refinability of the four point scheme. CAGD 19(4), pp235–238, 2002

[Prot02]   V.Yu.Protasov: On the Decay of Infinite Products of Trigonometric Polynomials. Mathematical Notes 72(6), pp819–832, 2002

[PSSK03]   P.Prusinkiewicz, F.Samavati, C.Smith and R.Karwowski: L-system description of subdivision curves. International Journal of Shape Modeling 9 pp41–59, 2003

[AHH]   R.Ait-Haddou and W.Herzog: Convex subdivision of a Bézier curve. CAGD 19(8), pp663–672, 2002

[HaDo03]   M.F.Hassan and N.A.Dodgson: Ternary and Three-point univariate subdivision schemes. pp 199–208 in Curve and Surface Fitting, (eds Cohen, Merrien and Schumaker), 2003

[Le03]   A.Levin: Polynomial generation and quasi-interpolation in stationary non-uniform subdivision. CAGD 20(1),pp41–60, 2003

[LeLe03]   A.Levin and D.Levin: Analysis of quasi-uniform subdivision. Applied and Computational Harmonic Analysis 15, pp18–32, 2003

[DLL03]   N.Dyn, D.Levin and A.Luzzato: Non-stationary interpolatory subdivision schemes reproducing spaces of exponential polynomials. Found Comput Math, pp197–206, 2003

[IDS04]   I.P.Ivrissimtzis, N.A.Dodgson and M.A.Sabin: A generative classification of mesh refinement rules with lattice transformations. CAGD 21(1) pp99–109, 2004

[ISD04]   I.Ivrissimtzis, M.A.Sabin, and N.A.Dodgson: On the support of recursive subdivision. ACM Trans. on Graphics 23(4) pp1043–1060, 2004

[DLM04]   N.Dyn, D.Levin and M.Marinov: Geometrically controlled 4-point interpolatory schemes. pp301–315 in Advances in Multiresolution for Geometric Modelling, (eds Dodgson, Floater, Sabin), Springer 2004, ISBN 3-540-21462-3

[Le05]   A Levin: The importance of polynomial reproduction in piecewise-uniform subdivision. pp272–307 in Mathematics of Surfaces XI, (eds Martin, Bez and Sabin) Springer LNCS3604 2005, ISBN 3-540-28225-9

[DFH05]   N.Dyn, M.S.Floater, K.Hormann: A C^2 four-point subdivision scheme with fourth order accuracy and its extensions. pp145–156 in Mathematical Methods for Curves and Surfaces, (eds Daehlen, Morken and schumaker), Nashboro Press 2005 ISBN 0-9728482-4-X

[SaDo05]   M.A.Sabin, and N.A.Dodgson: A circle-preserving variant of the four-point subdivision scheme. pp275–286 in Mathematical Methods for Curves and Surfaces, (eds Daehlen, Morken and Schumaker), Nashboro Press 2005 ISBN 0-9728482-4-X

[ChHa05]   D.Chen and M.Han: Convergence of cascade algorithm for individual initial function and arbitrary refinement masks. Science in China A 48(3) pp350–359, 2005

[MDL05]   M.Marinov, N.Dyn amd D.Levin: Geometrically controlled 4-point interpolatory scheme. pp301–315 in Advances in Multiresolution for Geometric Modelling (eds Dodgson, Floater, Sabin), Springer 2005 ISBN 3-540-21462-3

[WaDy05]   J.Wallner and N.Dyn: Convergence and C^1 analysis of subdivision schemes on manifolds by proximity. CAGD 22(7) pp 593–622, 2005

[Wa06]   J.Wallner: Smoothness Analysis of Subdivision Schemes by Proximity. Constr.Approx. 24 pp289–318, 2006

[Yang06]   X.Yang: Normal based subdivision scheme for curve design. CAGD 23(3), pp243–260, 2006

[CLLY06]   S.W.Choi, B.-G.Lee, Y.J.Lee and J.Yoon: Stationary subdivision schemes reproducing polynomials. CAGD 23(4), pp351–360, 2006

[BGR07a]   C.Beccari, G.Gasciola and L.Romani: A non-stationary uniform tension controlled interpolating 4-point scheme reproducing conics. CAGD 24(1), pp1–9, 2007

[BGR07b]   C.Beccari, G.Gasciola and L.Romani: An interpolating 4-point C^2 ternary non-stationary subdivision scheme with tension control. CAGD 23(4), pp210–219, 2007

[CDS07]   T.J.Cashman, N.A.Dodgson and M.A.Sabin: Non-uniform B-spline subdivision using refine and smooth. pp121–137 in Mathematics of Surfaces XII, (eds Martin, Sabin and Winkler), Springer LNCS4647, 2007 ISBN 3-540-73842-8

[BGR07]   C.Beccari, G.Gasciola and L.Romani: A Vector Outline Descriptor using interpolatory subdivision curves. pp31–40 in Curve and Surface Design, (eds Chenin, Lyche and Schumaker), Nashboro Press 2007, ISBN 978-0-9728482-7-5

[PeRe08]   J.Peters and U.Reif: Subdivision Surfaces. Springer 2008, ISBN 978-3-540-76405-2

[HoSa08]   K.Hormann and M.A.Sabin: A family of subdivision schemes with cubic precision, CAGD 25(1), pp 41–52, 2008

[ScWa08]   S.Schaefer and J.Warren: Exact evaluation of limits and tangents for non-polynomial subdivision schemes. CAGD 25(8), pp607–620, 2008

[SVG08]   S.Schaefer, E.Vouga and R.Goldman: Nonlinear subdivision through nonlinear averaging. CAGD 25(3) pp162–180, 2008

[HEMI09]   V.Hernandez–Mederos, J.C.Estrada–Sarlabous, S.R.Morales and I.Ivrissimtzis: Curve subdivision with arc–length control, Computing 86, pp151–169, 2009

[DeWa09]   CY.Deng and GZ.Wang: Generating planar spiral by geometry driven subdivision scheme. Science in China F 52(10), pp1821–1829, 2009

[DyOs09]   N.Dyn and P.Oswald: Univariate subdivision and multiscale transforms: The nonlinear case. pp203–247 in Multiscale, Nonlinear and Adaptive Approximation (eds R.DeVore, A.Kunoth), Springer, 2009

[Sa09]   M.Sabin: Two open questions relating to subdivision. Computing 86 pp213-217, 2009

[ScGo09]   S.Schaefer and R.Goldman Non-uniform subdivision for B-splines of arbitrary degree. CAGD 26(1), pp75–81, 2009

[CDS09a]   T.J.Cashman, N.A.Dodgson and M.A.Sabin   A symmetric, non-uniform, refine and smooth subdivision algorithm for general degree B-splines. CAGD 26(1), pp94–104, 2009

[CDS09b]   T.J.Cashman, N.A.Dodgson and M.A.Sabin: Selective knot insertion for symmetric, non-uniform, refine and smooth B-spline subdivision. CAGD 26(4), pp472–479, 2009

[DaSh09]   S.Daniel and P.Shunmugaraj; An approximating C^2 non-stationary subdivision scheme. CAGD 26(7), pp810–821, 2009

[DFH09]   N.Dyn, M.S.Floater and K.Hormann: Four-point curve subdivision based on iterated chordal and centripetal parameterizations. CAGD 26(3), pp279–286, 2009

[Ro10]   L.Romani: A circle-preserving C^2 Hermite interpolatory subdivision scheme with tension control. CAGD 27(1), pp36–47, 2010

[DW10]   C.Deng and G.Wang: Incenter subdivision scheme for curve interpolation. CAGD 27(1), pp48–59, 2010

[ADS10]   U.H.Augsdörfer, N.A.Dodgson and M.A.Sabin: Variations on the four-point subdivision scheme. CAGD 27(1), pp78–95, 2010

[CoMa10]   P.Costantini and C.Manni: A geometric approach for Hermite subdivision. Numer.Math. 2010

Index